Two Manuals Combined into One Book

Military Police Operations

And

Military Working Dogs

Department of Defense

Pentagon
Publishing

***FM 3-19.1** (FM 19-1)

Field Manual
No. 3-19.1

Headquarters
Department of the Army
Washington, DC, 22 March 2001

Military Police Operations

Contents

Distribution Restriction: Approved for public release; distribution is unlimited.

*This publication supersedes FM 19-1, 23 May 1988.

Preface

The Military Police (MP) Corps supports the commander across the full spectrum of military operations. This manual is the foundation for all MP doctrine as it relates to this support. It communicates to all levels of leadership and staffs how the MP provide a flexible and lethal force capable of operating across this full spectrum. As the keystone manual, it identifies what the MP train on and how their forces are organized and equipped in support of all Army echelons. Additionally, this manual provides guidance that can be used by United States (US) Army service schools, students, sister services, and federal agencies.

This manual is based on the purpose, organization, responsibilities, and goals of the US Army as set forth in Field Manuals (FMs) 100-1 and 3-0, as well as corps, division, and brigade manuals. Additionally, this manual is fully compatible with current joint, multinational, and interagency doctrine.

Appendix A contains a metric conversion chart.

The proponent of this publication is HQ TRADOC. Send comments and recommendations on Department of the Army (DA) Form 2028 directly to Commandant, US Army Military Police School (USAMPS), ATTN: ATSJ-MP-TD, 401 Engineer Loop, Suite 2060, Fort Leonard Wood, Missouri 65473-8926.

Unless this publication states otherwise, masculine nouns or pronouns do not refer exclusively to men.

Chapter 1

Introduction

The MP Corps has a strong history evolving over the past five decades. We, as a corps, continue to transform our organizations and doctrine as we have in the past to support the Army in the active defense of the 1970s, the AirLand battle of the 1980s, and now the force-projection doctrine of the 1990s. Our five MP functions clearly articulate the diverse role the MP play across the full spectrum of military operations. We cannot bask in our successes, nor reflect or celebrate. Our charter is to continue our legacy of stellar performance and strive to perfect it.

BG Donald J. Ryder

When the Army developed the Active Defense strategy in 1976, the US was facing the Cold War scenario of central Europe. Military strategy and doctrine were related to a single, focused threat that revolved around the countries in the Warsaw Pact. We were an outnumbered and technically inferior force facing an armor-dominated European battlefield. The MP Corps supported the Active Defense strategy by tailoring its forces to meet the threat. In 1982, when the AirLand Battle strategy was developed, US forces were still outnumbered, but were no longer technically inferior. Still threat-based and focused on a central European conflict, the AirLand Battle strategy used a relatively fixed framework suited to the echeloned attack of soviet-style forces. It delineated and clarified the levels of war; emphasized closed, concerted operations of airpower and ground forces; balanced the offense and the defense; and highlighted the synchronization of close, deep, and rear operations. MP doctrine kept pace with the Army's AirLand Battle strategy by supporting the battlefield commander through four basic missions—battlefield circulation and control, area security (AS), enemy prisoner of war (EPW), and law and order (L&O).

OVERVIEW

1-1. In October 1983, MP capabilities in the AirLand Battle strategy were tested during operation Urgent Fury in Grenada. The MP performed missions that ranged from assisting the infantry in building-clearance operations to assisting Caribbean peacekeeping forces in restoring L&O. These actions secured the MP's place in the combat-support (CS) role, demonstrating the professional knowledge and flexibility necessary for rapid transition from combat to CS to peacetime missions. The changing battlefield conditions of operation Urgent Fury set the stage for the demand of MP units today.

1-2. Evolving simultaneously with the changing definition of the modern battlefield, MP performance in Operations Hawkeye, Just Cause, and Desert Shield/Storm galvanized their ability to perform at any point along the operational continuum. With the publication of FM 100-5 in 1993, the Army adopted the doctrine of full-dimensional operations, relying on the art of battle command to apply those principles and to shift the focus from AirLand Battle to force-projection doctrine. This new doctrine was based on recent combat experience in a multipolar world with new technological advances. Already trained and expected to perform in this new strategy, MP support was already in place and fully operational. The MP continued to perform their basic battlefield missions and to refine their capabilities while supporting the battlefield commander as he deployed to contingency operations throughout the world.

1-3. In 1996, the MP Corps went through a doctrinal review process to determine if it was properly articulating its multiple performance capabilities in support of US forces deployed worldwide (see Appendix B). The review process identified the need to restructure and expand the EPW mission to include handling US military prisoners and all dislocated civilians. This new emphasis transformed the EPW mission into the internment and resettlement (I/R) function. The review process also identified the need to shift from missions to functions. In the past, the four battlefield missions adequately described MP capabilities in a mature theater against a predictable, echeloned threat. However, that landscape is no longer valid. Accordingly, the four MP battlefield missions have become the following five MP functions:

- Maneuver and mobility support (MMS).
- AS.
- L&O.
- I/R.
- Police intelligence operations (PIO).

1-4. These new MP functions are shaped by the following factors:

- The application of stability and support operations where the integration of joint, multinational, and interagency capabilities are common occurrence.
- The lack of traditional linear battlefields, requiring theater commanders in chief (CINCs) to request forces that meet a specific function to accomplish operational requirements.
- The impact of asymmetric threats (such as drug traffickers and terrorist factions) and the effects of man-made and natural disasters.
- The impact of advances in information and communication technologies and specifically in understanding the increased vulnerabilities presented by these technologies.

1-5. Articulating MP capabilities along functional lines benefits the MP and the Army echelon commander as well as the combatant commander. Since there is a multinational, interagency, and sister-service overlap of security services, the importance of including MP leaders and staffs early in the operational planning process cannot be overemphasized. This means before units are designated, unit boundaries are drawn, and unit missions are assigned. Early involvement ensures the proper development of common security responsibilities, communication and connectivity, liaisons, processes, and the rules of interaction between all forces. The ultimate goal should be the optimal, phased employment of MP forces in support of a commander's operational plan. MP functions not only reflect and capture current capabilities, they define the MP Corps in the twenty-first century.

1-6. As the Army reshapes and focuses its resources on transformation, Force XXI, and other redesign efforts, the MP Corps stands proud and ready to support this progress and reiterate its commitment to assist, protect, and defend.

OPERATIONAL FRAMEWORK

1-7. The operational framework consists of the arrangement of friendly forces and resources in time, space, and purpose with respect to each other, the enemy, or the situation (see Figure 1-1). The operational framework for Army forces (ARFOR) rests within the combatant commander's theater organization. Each combatant commander has an assigned geographical area of responsibility (AOR), also called a theater, within which he has the authority to plan and conduct operations. Within the theater, joint-force commanders at all levels may establish subordinate operational areas such as areas of operation (AOs), joint-operations areas (JOAs) and joint rear areas

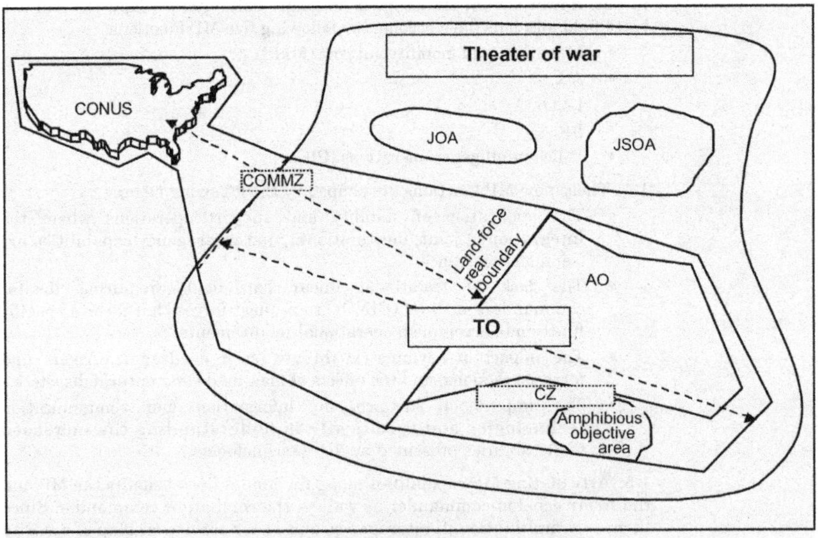

Figure 1-1. Operational Framework

(JRAs). The JRAs facilitate the protection and operation of bases, installations, and forces that support combat operations. When warranted, combatant commanders may designate theaters of war, theaters of operations (TOs), combat zones (CZs), and communications zones (COMMZs).

1-8. A theater of war is that area of air, land, or water that is, or may become, directly involved in the conduct of the war. A theater of war may contain more than one TO. It does not normally encompass the geographic combatant commander's entire AOR. A TO is a subarea (defined by a geographic combatant commander) within a theater of war in which specific combat operations are conducted or supported.

1-9. A CZ is the area required by combat forces for conducting operations. It normally extends forward from the land force's rear boundary. The COMMZ is the rear part of the TO (behind but contiguous to the CZ) that contains the lines of communication (LOC) and provides supply and evacuation support. Other agencies required for the immediate support and maintenance of field forces may also be located in the COMMZ. The COMMZ spans back to the continental US (CONUS) base, to a supporting combatant commander's AOR, or both.

1-10. An AO is an operational area defined by the joint-force commander for land and naval forces. An AO does not typically encompass the entire operational area of the joint-force command (JFC), but it should be large enough for component commanders to accomplish their mission and protect their forces. Army commanders use control measures to describe AOs and to design them to fit the situation and take advantage of the joint force's capabilities. Commanders typically subdivide the assigned AO by assigning subordinate-unit areas. These subordinate-unit areas may be contiguous or noncontiguous (see Figure 1-2). When friendly forces are contiguous, a boundary separates them. When friendly forces are noncontiguous, the concept of operations links the force's elements, but the AOs do not share a boundary. The intervening area between noncontiguous AOs remains the responsibility of the higher headquarters.

BATTLEFIELD ORGANIZATION

1-11. Battlefield organization is the arrangement of forces according to purpose, time, and space to accomplish a mission. Battlefield organization has both a purpose- and spatial-based framework. The purpose-based framework centers on decisive, shaping, and sustaining (DSS) operations. Purpose unifies all elements of the battlefield organization by providing the common focus for all actions. However, forces act in time and space to accomplish a purpose. The spatial-based framework includes close, deep, and rear areas. Despite the increasing nonlinear nature of operations, there may be situations where commanders describe DSS operations in spatial terms. Typically, linear operations involve conventional combat and concentrated maneuver forces. Ground forces share boundaries and orient against a similarly organized enemy force. In such situations, commanders direct and focus simultaneous DSS operations in deep, close, and rear areas, respectively (see FM 3-0).

1-12. The MP battlefield organization supports every Army echelon, from the Army service component command (ASCC) and the theater-support command

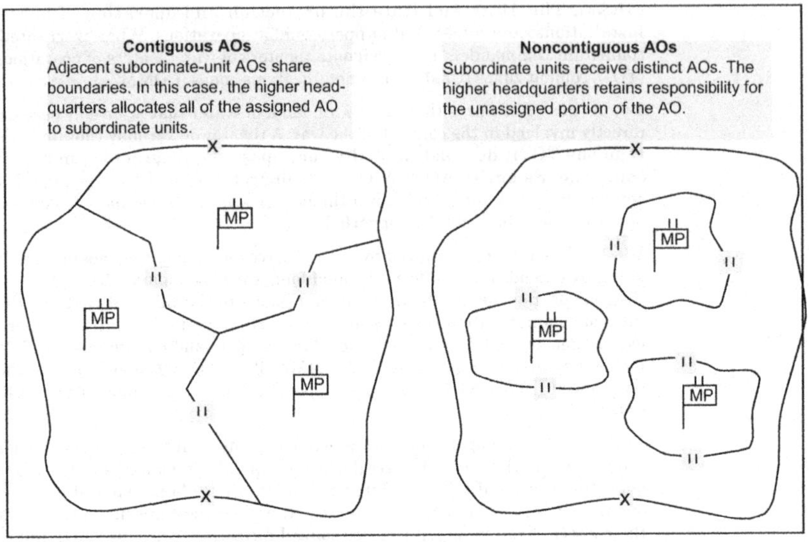

Figure 1-2. Contiguous and Noncontiguous AOs

(TSC) to the maneuver brigade. Regardless of the battlefield organization (purpose or spatial based), MP support to the Army commander is based on available resources and mission, enemy, terrain, troops, time available, and civilian considerations (METT-TC).

1-13. MP support throughout the theater of war may include MP units in the JOA and in the TO. If the combatant commander designates a COMMZ and a CZ within his TO, MP support will come from the established MP modified table of organization and equipment (MTOE) at the subordinate-command echelon. MP support to the JOA is also provided based on METT-TC and available MP assets. Typical MP support may include an I/R brigade liaison detachment (BLD), MP brigades and battalions, a division MP company, a military-working-dog (MWD) team, an L&O team, and a customs team. Figure 1-3, page 1-6, depicts a typical MP organization throughout the TO. In the COMMZ, Figure 1-3 depicts the different types of MP units that are assigned to echelons above corps (EAC) (the ASCC or the TSC). In the CZ, Figure 1-3 depicts the different types of MP units that are assigned to corps, division, and the separate brigades.

SUPPORT IN THE COMMZ

1-14. MP support in the COMMZ is provided by an array of multifunctional MP units. The following MP units provide MP support to EAC:

- The MP brigade (I/R). The MP brigade (I/R) may augment the ASCC or the TSC during wartime. Its mission is to provide command, staff

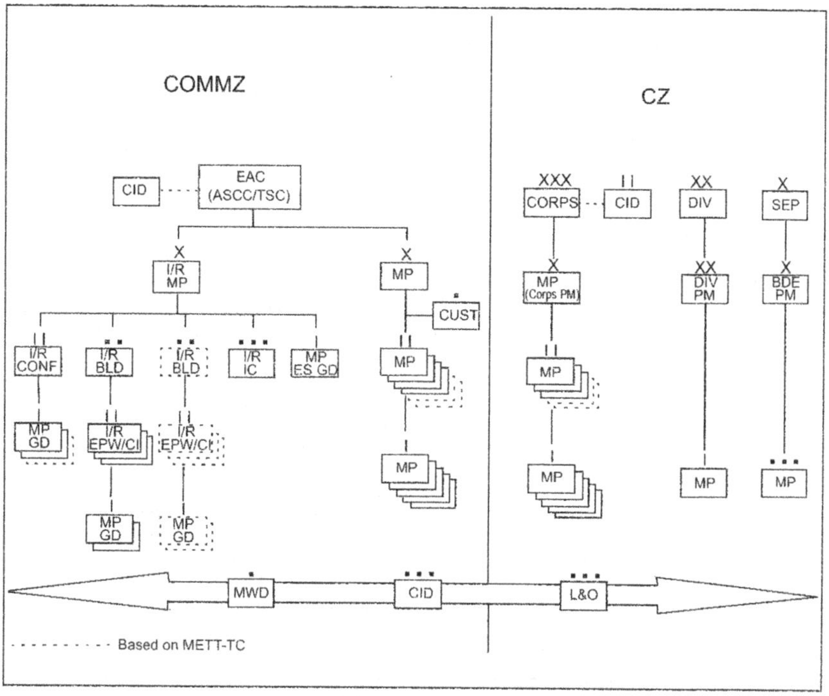

Figure 1-3. MP Structure in the TO

planning, and supervision of I/R operations. This includes coordination with joint and host-nation (HN) agencies, civilian police authority, nongovernmental organizations (NGOs), and US federal agencies.

- The MP brigade (CS). The MP brigade (CS) is assigned to the ASCC or the TSC during wartime (based on METT-TC). The MP brigade (CS) is capable of performing all five MP functions.
- The Criminal Investigation Division (CID) group. The CID group is a stovepipe organization that reports directly to the Commander, US Army Criminal Investigation Command (USACIDC). The CID group provides support to the ASCC and subordinate commands (TSC, corps, or division). See Chapter 9 for further discussion of CID support.

1-15. MP support to other EAC subordinate commands is performed only if MP resources are available. See Chapter 5 for further discussion of MP support to EAC.

SUPPORT IN THE CZ

1-16. MP support is provided in the CZ to each corps, division, and brigade (separate teams or initial/interim brigade combat teams [IBCTs]). An MP brigade (CS) is assigned to each corps, and the MP brigade commander is the corps's provost marshal (PM). A PM and his section, along with an organic division MP company, are assigned to each division. A PM cell and an MP platoon are organic to a separate brigade. A two-person PM cell is organic to the IBCT. The MP units assigned to corps, divisions, and separate brigades are capable of performing all five MP functions. They provide combat, CS, and combat-service-support (CSS) operations within their command's AO.

TYPES OF MILITARY POLICE UNITS

1-17. Most MP units supporting a TO and a JOA are capable of performing all five MP functions. However, the functions must be prioritized based on METT-TC and the availability of MP assets. Current MP structures are designed and tailored to better support the level of command deployed. For example, at the division level, division MP companies are organized as light, heavy, airborne, or air assault and are organic to their respective divisions. The EAC and corps MP brigades and battalions are equally designed to command and control a force mix of up to six battalions or companies. An MP escort-guard and guard company are designed to transport, guard, and provide security to EPWs, civilian internees (CIs), or dislocated civilians. The MP escort-guard company is assigned to the MP brigade (I/R), and the MP guard company is assigned to the MP battalion (I/R).

1-18. MP units can also be tailored and augmented to accomplish multiple, diverse, or specific missions. Customs, L&O, and MWD teams are examples of MP capabilities and flexible responses to a combatant commander's operational requirements. (See Table 1-1, pages 1-9 through 1-19, for a more complete description of MP units. See FM 19-10 and Army Regulation (AR) 190-12 for further information.) The battlefield workload analysis (BWA) is a tool used to determine the number of MP units required to perform some of these multiple missions (see Appendix C).

JOINT, MULTINATIONAL, AND INTERAGENCY OPERATIONS

1-19. In today's environment, the Army will rarely operate or fight alone. The high probability that the Army will operate in concert with its sister services, in an alliance with the forces of foreign nations, or in support of United Nations (UN) operations (when it is committed) is fully reflected in joint doctrine. In such operations, protecting LOC, key facilities, and command and control (C^2) centers will be a shared responsibility. Under this framework, MP units can expect to share the AO with joint, combined, multinational, or interagency resources. MP forces must be prepared to conduct a number of full-spectrum operations with a variety of government and nongovernmental agencies, other services, allied nations, and international agencies.

1-20. Corps and division commanders and staffs must plan (in advance) the transition from a single-service headquarters with joint representation to a joint headquarters capable of functioning as a joint task force (JTF) headquarters. When tasked to form a JTF headquarters, the corps or division must ensure that all of the staff sections and agencies have joint representation (see FMs 100-15 and 71-100). To this end, MP planners must ensure that the JTF is augmented with the appropriate MP forces and with the appropriate echeloned C^2.

1-21. Regardless of the force mix, the MP provide the force with unparalleled, multifunctional capabilities. Among these capabilities is the MP's ability to generate firepower or to handle populations such as EPWs/CIs, dislocated civilians, and refugees. Additionally, MP expertise in investigations and law enforcement enhances the capabilities of other joint, multinational, and interagency police and security forces.

1-22. MP security plans must reflect the joint synergy derived from combining the multiple and diverse capabilities of all participants. To capitalize upon that synergy, MP leaders must keep an open line of communication and coordination to offset the challenges presented by interoperability. Some of these interoperability challenges include—

- Differing political objectives.
- Differing capabilities.
- Cultural/language differences.
- Legal and policy constraints.
- Media impacts.
- Compromise of sensitive processes, procedures, and equipment.
- C^2.
- Communications (digital- and analog-equipment differences).

1-23. MP plans must also accommodate differences in planning capabilities, as well as differences in doctrine, training, and equipment. The intent is to match security missions with force capabilities. MP leaders must understand that operations will often involve multinational teams. While US forces routinely task-organize, this may be more difficult to accomplish with some multinational security forces. This kind of orchestration requires employing standardized procedures, communications, equipment, and liaison within the constraints of operations security (OPSEC).

1-24. Coordination is the key to mission accomplishment in multinational and interagency operations. A military coordination center or a civil-military operations center (CMOC) may meet this coordination requirement. The CMOC provides access for nonmilitary agencies desiring military (to include MP/CID) assistance and coordination. These nonmilitary agencies may include—

- Government organizations (GOs).
- NGOs.
- International organizations (IOs).
- International humanitarian organizations (IHOs).
- HN authorities and agencies.

1-25. The introduction of US Army MP forces in any joint, multinational, or interagency operation is based on METT-TC and the capabilities they bring to the operation. Effective integration of MP forces with other security forces reduces redundant functions, clarifies responsibilities, and conserves resources.

Table 1-1. Unit Descriptions

MP Unit	Mission	Action Elements	Support Role	Capabilities (Include But Are Not Limited to [See TOEs for Details])	Augmentation Needed for Mission Accomplishment
DIVISION MP UNITS					
MP Company (Airborne Division)	Perform the 5 MP functions.	A PM cell and 4 platoons. Each platoon has 3 squads and each squad has 2 three-man teams. Total: 24 three-man teams.	1 platoon deploys with each division brigade and provides DS during the assault phase. Then platoons revert to GS.	During the assault phase, 6 mounted, mobile or fixed-position teams deploy to provide DS for each brigade and 6 teams deploy to provide GS near the division main CP OR after the assault phase, 1 division EPW collection point (6 teams) and 18 mounted, mobile patrols or fixed-position teams provide GS OR 24 mounted, mobile or fixed-position teams are used in any combination.	The division band to provide close-in security for the division CP or to guard EPWs AND the corps CS MP to conduct combat operations within the division rear and provide sustained MMS, area security, special operations support, and escort for evacuating EPWs/CIs between division collection points and to the corps holding area AND the corps L&O MP and CID to conduct sustained L&O operations and criminal investigations AND division/corps transportation assets for evacuation of EPWs **NOTE: Both heavy division companies require corps CS MP augmentation for each forward support platoon (1 corps MP squad per forward support platoon).**
MP Company (Light Infantry Division)	Perform the 5 MP functions.	A PM cell and 3 platoons. Each platoon has 3 squads and each squad has 2 three-man teams. Total: 18 three-man teams	Provide GS for all units in the AO.	Provide screening security for the division main CP (6 teams), the division EPW collection point (6 teams), and 6 mounted, mobile or fixed-position teams.	
MP Company (Heavy Division)* *19333F000	Perform the 5 MP functions.	A PM cell and 2 division support platoons. Each platoon has 3 squads and each squad has a squad leader and 3 three-man teams AND 3 forward support platoons. Each platoon has 2 squads and each squad has a squad leader and 3 three-man teams. Total: 36 three-man teams	2 platoons provide GS and 3 platoons provide DS.	Provide screening security for the division main CP (6 teams), the division EPW collection point (6 teams), and 6 mounted, mobile or fixed-position teams AND EITHER 3 DS platoons providing their brigade with either 1 forward EPW collection point (3 teams) and 3 mounted, mobile or fixed-position teams or 6 teams in any combination OR 18 mounted, mobile or fixed-position teams are used in any combination.	
MP Company (Heavy Division)* *19333L000	Perform the 5 MP functions.	A PM cell and 5 platoons. Each platoon has 2 squads and each squad has 3 three-man teams. Total: 36 three-man teams	3 platoons provide GS and 3 platoons provide DS.	Provide screening security for the division main CP (6 teams), the division EPW collection point (6 teams), and 6 mounted, mobile or fixed-position teams AND EITHER 3 DS platoons providing their brigade with either 1 forward EPW collection point (3 teams) and 3 mounted, mobile or fixed-position teams or 6 teams in any combination OR 18 mounted, mobile or fixed-position teams are used in any combination.	

Table 1-1. Unit Descriptions (continued)

MP Unit	Mission	Action Elements	Support Role	Capabilities (Include But Are Not Limited to [See TOEs for Details])	Augmentation Needed for Mission Accomplishment
MP Company (Air Assault Division)	Perform the 5 MP functions.	A PM cell and 4 platoons. Each platoon has 3 squads and each squad has 2 three-man teams. Total: 24 three-man teams	Provide GS for all units in the AO.	24 mounted, mobile or fixed-position teams are used in any combination.	(See augmentation needs on previous page.)
MP COMBAT SUPPORT UNITS					
HHC MP Brigade	Provide C2 and coordinate the combat, CS, and CSS of all MP elements assigned or attached.	Command and staff elements and long-range planning section	Provide command, control, planning, and supervision for up to 6 MP battalions and other assigned or attached subordinate HQ.	Provide continuous command, control, coordination, planning, and supervision for subordinate units.	None
HHD MP Battalion (Combat Support)	Provide C2 and coordinate the combat, CS, and CSS of all MP elements assigned or attached.	Command and staff elements plus a support element	Provide command, control, planning, and supervision for up to 6 MP companies and other assigned or attached subordinate HQ.	Provide continuous command, control, coordination, planning, and supervision for subordinate units.	None
MP Company (Combat Support)	Perform the 5 MP functions.	4 platoons with 3 squads. Each squad has a squad leader and 3 three-man teams. Total: 36 three-man teams	Provide GS for all units in the AO.	Provide security for a unified or combined HQ or higher HQ or 1 main CP and 1 tactical CP OR operate 1 EPW holding area (9 teams) and 27 mounted, mobile or fixed-position teams for use in any combination.	Corps band assets to guard EPWs AND corps transportation for evacuation of EPWs AND external MP to escort EPWs
MP Company (Combat Support)	Perform the 5 MP functions.	4 platoons with 3 squads. Each squad has a squad leader and 3 three-man teams. Total: 36 three-man teams	Provide GS for all units in the AO.	Provide security for a unified or combined HQ or higher HQ or 1 main CP and 1 tactical CP OR operate 1 EPW holding area (9 teams) and 27 mounted, mobile or fixed-position teams for use in any combination.	Corps band assets to guard EPWs AND corps transportation for evacuation of EPWs AND external MP to escort EPWs
MP Company Arctic Support Augmentation Detachment	Augment MP units when operating in an arctic environment.	1 mechanic (63B10)	Provide mobility support to MP units.	Provide support as designated by the commander.	None
MP Detachment (C2)	Provide command and administrative personnel for MP law enforcement teams.	19503LA has a platoon HQS; 3 individuals. 19503LD has a commander and support personnel; 8 individuals.	Provide L&O. 19503LA provides platoon-level C2. 19503LD provides company-level command, control, and support.	Commands and controls law enforcement teams, investigations teams, MWD teams, and physical-security teams.	None

Table 1-1. Unit Descriptions (continued)

MP Unit	Mission	Action Elements	Support Role	Capabilities (Include But Are Not Limited to [See TOEs for Details])	Augmentation Needed for Mission Accomplishment
MP Detachment (Old Guard)	Perform the 5 MP functions.	4 squads. Each squad has 3 three-man teams. Total: 12 teams	Support the HHC old guard battalion.	Provide support as directed by the commander.	None
MP LAW AND ORDER (L&O) UNITS					
MP HQ Team	Provide C2, personnel administration, and logistical support to attached L&O augmentation teams.	Commander and support personnel	Provide L&O.	Provide C2, personnel administration, and logistical support to attached L&O augmentation teams.	None
MP Operations Team	Provide mission supervision, staff planning, and technical supervision required to support the L&O mission.	L&O operations officer and operations staff	Provide L&O.	Provide overall mission supervision, planning, employment, and coordination of support. Provide technical supervision and administrative support of MPI, TAI, and force protection MP. Provide an evidence custodian/police intelligence NCO to be responsible for confiscated property used in criminal offenses and to receive and collect the initial analysis of criminal/operational information/intelligence.	None
MP Desk or Desk and Record Team	Serve as the primary control point for L&O-related incidents and operations initiated by MP patrols.	2 three-man teams	Provide L&O.	Receive/record complaints, dispatch investigative persons, maintain control of offenders/detained persons, maintain the status of investigations, prepare reports, and provide criminal data to the police intelligence NCO for analysis.	None
MP Traffic Accident Investigations Team	Provide the technical capability to investigate traffic accidents.	1 two- or three-man team: 19517AD00 has 1 two-man team. 19523LE00 has 1 three-man team.	Provide L&O.	Augment MP capabilities to enforce MSR regulations; man checkpoints, roadblocks, and dismount points; patrol traffic areas; and perform traffic escort duties.	The 19517AD00 accident investigation team is dependent on the supported unit for transportation.
MP Investigation Team	Provide the technical capability to investigate criminal incidents and conduct surveillance operations.	One, two, or three-man teams: 19517AE has 1 two-man team. 19533LA has 1 one-man team. 19533LB has 1 one-man team. 19533LC has 1 three-man team.	Provide L&O.	Investigate criminal incidents, conduct surveillance, work with HN military and civilian police, and collect criminal and operational intelligence. 19533LA00 provides supervision for the teams. 19533LB00 provides investigative support for populations of not less than 700 troops. 19533LC00 provides investigative support for populations of not less than 2,100 troops.	Dependent on the supported unit for transportation

Table 1-1. Unit Descriptions (continued)

MP Unit	Mission	Action Elements	Support Role	Capabilities (Include But Are Not Limited to [See TOEs for Details])	Augmentation Needed for Mission Accomplishment
MP Force Protection and Physical-Security Teams	Provide the expertise and technical capability to assist units in safeguarding personnel, equipment, and facilities.	One, two, or three-man teams: 19517AF has 1 two-man team. 19593LA has 1 three-man team. 19593LB has 1 one-man team.	Provide L&O and area security.	Conduct vulnerability assessments and identify and prioritize critical facilities and key terrain within the AO. Identify mission-essential activities that are vulnerable to criminal acts or disruptive activities. Ensure that activities are inspected to determine if safeguards are adequate. Assist base/base cluster commanders with the development of internal defense plans.	The 19517AF is dependent on the 19517AB operations team for transportation.
MP Detachment (Patrol Supervision Team)	Provide supervision for 3 MP teams.	1 MP NCO	Provide L&O.	Plan, direct, and supervise the employment of assigned or attached MP teams.	Dependent on the supported unit for transportation
MP Detachment (Motor and Dismounted Patrol Team)	Perform L&O operations.	1 three-man team	Provide L&O.	Protect designated personnel or facilities. Perform route and area reconnaissance, enforce MSR regulations, provide refugee and straggler control, and disseminate information. Perform peacekeeping operations to maintain surveillance over an area, observe activities, and report findings. Preserve or establish L&O. Apprehend absentees or deserters (US military personnel) in conjunction with civil law enforcement agencies.	The dismounted patrol team is dependent on the supported unit for transportation, when required.
MP Detachment (Registration Team)	Perform vehicle and firearm registration.	1 three-man team	Provide L&O.	Service up to 10,000 personnel for registration of individuals, vehicles, and firearms on a closed post or area. Issue credentials prescribing limits of circulation and privileges. Process up to 50 personnel daily for fingerprints and photography services.	Dependent on the supported unit for transportation
Senior Military Customs Inspector Supervisor	Perform technical supervision, staff planning, and coordination for customs inspection teams.	1 MP NCO	Provide L&O.	Supervise two senior military customs teams to ensure that personnel, equipment, and material meet customs, immigration, Department of Agriculture, and other federal agency requirements for units and personnel redeploying to the US.	None

Table 1-1. Unit Descriptions (continued)

MP Unit	Mission	Action Elements	Support Role	Capabilities (Include But Are Not Limited to [See TOEs for Details])	Augmentation Needed for Mission Accomplishment
Senior Military Customs Inspection Team	Conduct customs inspections and train and supervise redeploying unit personnel to augment US customs inspectors.	1 five-man team	Provide L&O.	Conduct inspections and advise redeploying units. Train selected individuals from the redeploying unit to augment US customs efforts. NOTE: One team supports one port of embarkation.	The redeploying unit undergoing customs inspection, to augment US Customs efforts. The number of personnel is proportional to the size of the unit.
MILITARY WORKING DOG (MWD) UNITS					
MWD Kennel Master Team	Supervise MWD teams.	One- or two-man team. 19537AA has 1 two-man team. 19583LG has 1 one-man team.	Provide L&O, area security, and I/R operations support.	Supervise, plan, and coordinate MWD operations and support requirements. Provide MWD handler/ canine proficiency certification.	None
MWD Explosive, Narcotics, and Patrol Team	Detect explosives and controlled substances and search for, detect, and control personnel in support of crime scene searches. Provide personal protection, MOUT, health and welfare inspections, and I/R and customs operations.	One-man/one-dog team. The 19537A000-series TOEs have 3 teams each. The 19583L000-series TOEs have 1 team each.	Provide L&O, area security, and I/R support.	The 19537A-series dog teams are capable of providing a 24-hour explosive, narcotics, and tracking handler/dog requirement or three concurrent short-duration missions each requiring 1 handler/dog team. The duration of the missions will vary based on the climate, the environment, and the individual dog's ability. The 19583L-series dog teams provide short- duration missions based on the dog's ability. Explosive and narcotics teams provide MWD patrol support when not employed with explosive and narcotics detection.	None
MILITARY POLICE INTERNMENT/RESETTLEMENT (I/R) UNITS					
MP Command (I/R)	Provide command, control, staff planning, and supervision of I/R operations performed by all assigned or attached elements.	Command and staff elements	Provide I/R operations support.	Provide command, control, and staff planning for 2 or more MP brigades.	MP IRIC AND MP EPW evacuation detachment

Table 1-1. Unit Descriptions (continued)

MP Unit	Mission	Action Elements	Support Role	Capabilities (Include But Are Not Limited to [See TOEs for Details])	Augmentation Needed for Mission Accomplishment
MP Brigade (I/R)	Provide command, control, and staff planning for I/R operations performed by all assigned or attached MP elements.	Command and staff elements	Provide I/R operations support.	Provide command, control, and staff planning for 2 to 7 MP I/R battalions or up to 21 MP I/R battalions when augmented by 7 brigade liaison detachments (BLDs). Plan and provide staff supervision of I/R collection and evacuation operations. Coordinate with HN military territorial organizations and civilian police authorities, nongovernment organizations, private volunteer organizations, and US federal agencies on I/R matters. Provide coordination and support for out-of-theater evacuation of EPWs, if required. Support posthostilities operations.	MP IRIC AND MP EPW evacuation detachment AND MP I/R BLD (TOE 19453LD00) **NOTE: The BLD expands the staff planning and coordination capabilities on a ratio of 1 BLD to 3 MP I/R battalions.**
MP Battalion (I/R)	Provide command, staff planning, administration, and logistical support for the operation of an I/R facility.	Command and staff elements	Provide I/R operations support.	Operate an internment facility for either EPWs/CIs or DCs or a confinement facility for US prisoners. (Never more that one category at the same time.)	MP detachment (I/R) (EPW/CI) for EPW/CI/DC internment missions AND MP detachment (I/R) (confinement) for US prisoner confinement missions AND MP company (EG) for security of the evacuation and/or movement of US prisoners or EPWs/CIs/DCs AND MP company (guard) for prisoner guard services AND Corps/EAC transportation assets for movement of EPWs/CIs/DCs

Table 1-1. Unit Descriptions (continued)

MP Unit	Mission	Action Elements	Support Role	Capabilities (Include But Are Not Limited to [See TOEs for Details])	Augmentation Needed for Mission Accomplishment
MP Detachment (I/R) (EPW/CI)	Augment the MP battalion (I/R) to provide supervision, administration, combat health support, and logistical support for operating an internment facility interning EPWs/CIs or housing DCs.	2 compound control sections (2 three-man teams), a work project section (1 two-man team), a personnel section, and a supply section.	Provide I/R operations support.	Provide command, control, and supervision of up to 1,000 EPWs/CIs or 2,000 DCs. Provide supply and subsistence support (to include supervision of food preparation) for up to 1,000 EPWs/CIs or 2,000 DCs. Direct the activities related to assigning and supervising work projects for EPWs/CIs. Augment the MP I/R battalion in processing and maintaining records for up to 1,000 EPWs/CIs or 2,000 DCs.	MP company (guard) for prisoner guard services
MP Company (Escort Guard)	Provide supervisory and security personnel for evacuating and/ or moving EPWs/CIs.	4 platoons with 3 squads each. Each squad has a squad leader and 3 three-man teams. Total: 36 teams	Provide I/R operations support.	Provide security for the movement of the following numbers of POWs or CIs by the methods indicated: Marching - 1,000 to 1,500. Vehicle - 1,500 to 2,000. Rail - 2,000 to 3,000. Air - 2,125 to 2,406.	Corps/EAC transportation assets for movement of EPWs/CIs
MP Company (Guard)	Provide guards for EPWs/CIs or US prisoners, installations, and facilities.	3 platoons with 3 squads each. Each squad has a squad leader and 2 five-man teams Total: 18 teams.	Provide I/R operations support.	Provide security for a confinement facility containing up to 500 US military prisoners. Provide guards for securing 1 EPW/CI compound containing up to 2,000 EPWs/CIs or 4,000 DCs. Provide security guards for 3 railway terminals, each having up to 8 tracks. (Includes EPW/CI rail movement operations and protection of sensitive material within the terminal.) Provide guards for 1 military installation or facility up to 240,000 square yards in size, containing sensitive material. (This unit can provide guards for material transit.)	None
MP I/R Processing Squad	Process EPWs/ CIs.	1 squad leader and 1 eight-man processing squad	Provide I/R operations support. Augment the I/R battalion processing capability to meet operational requirements.	Provide processing capabilities of about 8 EPWs/CIs per hour.	None

Table 1-1. Unit Descriptions (continued)

MP Unit	Mission	Action Elements	Support Role	Capabilities (Include But Are Not Limited to [See TOEs for Details])	Augmentation Needed for Mission Accomplishment
I/R Brigade Liaison	Expand the MP brigade (I/R) or ASCC C2 capabilities.	Liaison officer and support staff	Provide I/R operations support.	When assigned to the MP I/R brigade, provide staff augmentation that expands the brigade's staff planning, coordination, and C2 capabilities for 3 MP I/R battalions. When assigned to the ASCC, provide I/R staff augmentation and a liaison link with allied/HN forces to ensure that the care and handling of US captured EPWs/CIs is in compliance with the Geneva Conventions.	None
I/R Camp Liaison Team	Provide continuous accountability of EPWs/CIs captured by US forces and transferred to an allied/HN for internment.	Liaison officer and support staff	Provide I/R operations support.	Provide advice, as requested, to commanders and staffs of allied/HN-operated internment facilities. Verify arrival, forward records, and provide continuous accountability for US captured EPWs/CIs interned in allied/HN facilities. Monitor prisoner treatment to ensure compliance with the Geneva Conventions. Receive/certify allied/HN requests for reimbursement of expenses associated with interning EPWs/CIs captured by US forces.	None
I/R Processing Liaison Team	Process and verify the transfer of EPWs/CIs captured by US forces to a allied/HN facility.	Liaison officer and support staff	Provide I/R operations support.	Provide processing and transfer capability for US-captured prisoners to an allied/HN EPW/CI facility.	None
MP Detachment (EPW Evacuation)	Provide evacuation support of EPWs.	EPW coordination officer and staff plus 10 one-man POD/POE teams and 3 two-man POE teams	Provide I/R operations support.	Coordinate administrative, logistical, and transportation support for up to 7 PODs and 6 POEs used for out-of-theater EPW evacuation and the escort guard personnel that support the movements. Coordinate for theater security at PODs/POEs and intermediate refuel points.	Responsible POD/POE security forces

Table 1-1. Unit Descriptions (continued)

MP Unit	Mission	Action Elements	Support Role	Capabilities (Include But Are Not Limited to [See TOEs for Details])	Augmentation Needed for Mission Accomplishment
MP Detachment (Internment Resettlement Information Cener [IRIC])	Provide a US central agency in each theater of war to receive, process, maintain, disseminate, and transmit the required information and data relating to EPWs, American POWs, CIs, and DCs within the theater.	Command and staff personnel	Provide I/R operations support. Serve as the single source for collection and storage of EPW/CI information in theater. Forward information to the National Prisoner of War Information Center at DA.	Collect, process, and disseminate (to authorized agencies) information regarding interned personnel detained in theater and those released to the custody of allied/HN authorities within theater. Receive, document, and disseminate to the theater commander information received regarding American POWs, CIs, and foreign nationals who are captured, missing, or otherwise detained. Receive, store, and dispose of personal property belonging to interned personnel who have died, escaped, or been repatriated and any property belonging to enemy soldiers killed in action that is not disposed of through grave registration channels.	None
MP Detachment (I/R) (Confinement)	Augment an MP battalion (I/R) in operating a confinement facility. Provide supervision, administration, combat health support, and logistical support of US military prisoners.	Corrections officer and confinement facility staff	Provide I/R operations support.	Provides a control team for the command, control, and supervision of US military prisoners and a disciplinary guard team to supervise custodial personnel for a maximum custody of 500 prisoners. Provide administration, health service, supply, and food service personnel to augment the battalion in providing these services for internees. (The food service personnel supervise and train US military prisoners working in the internee dining facility.)	MP battalion (I/R) for command, staff, planning, and operational support AND MP company (guard) for prisoner guard services

Table 1-1. Unit Descriptions (continued)

MP Unit	Mission	Action Elements	Support Role	Capabilities (Include But Are Not Limited to [See TOEs for Details])	Augmentation Needed for Mission Accomplishment
MP Confinement Facility Detachment	Provide C2, staff planning, administration, and logistical support for a confinement facility for US military prisoners.	Commander and facility staff	Provide I/R operations support.	Provide command, control, and administrative support for a confinement facility.	MP guard company to provide exterior perimeter security and required guard support to the confinement facility AND MP correctional team supervisor (TOE19553LE) to provide supervision of the guard force AND MP correctional team work supervisor (TOE 19553LF) to provide supervision of prisoners
MP Correctional Team Supervisor	Provide correctional supervision of prisoner work guard forces within a confinement facility.	2 two-man teams	Provide I/R operations support.	Provide correctional supervision to guard force personnel and US military prisoners within a modular-configured confinement facility.	None
MP Correctional Team Work Supervisor	Provide direct control and supervision of military prisoners within a confinement facility.	1 two-man team	Provide I/R operation support.	Provide correctional control, supervisory functions, and escort duties for military prisoners within a modular-configured confinement facility.	None
MILITARY POLICE CRIMINAL INVESTIGATION (CID) UNITS					
MP Group (CID)	Provide command, control, staff planning, and supervision for all CID elements within the theater.	Command and staff elements	Provide L&O.	Provide command, control, staff planning, and supervision over all CID elements within the theater. **NOTE: The Criminal Investigation Command is a stovepipe organization.**	None

Table 1-1. Unit Descriptions (continued)

MP Unit	Mission	Action Elements	Support Role	Capabilities (Include But Are Not Limited to [See TOEs for Details])	Augmentation Needed for Mission Accomplishment
MP Battalion (CID)	Provide command, control, staff planning, and supervision for all assigned or attached CID elements.	Command and staff elements	Provide L&O.	Provides command, control, staff planning, and supervision over all assigned and attached CID elements.	None
MP Detachment (CID) HQ Cell	Provide C2, evidence custody control, and investigative administration support. Coordinate for personnel administration and logistic support.	Commander and support staff	Provide L&O.	Provide C2 of assigned CID SA sections and/or teams performing all CID operations in their area of responsibility.	None
MP Detach-ment (CID)	Provide criminal investigative support to Army commanders at all echelons.	2-man teams. The number of teams varies by organization and is as follows: DSE heavy - 4 DSE light - 3 Section A - 4 Section B - 4 Supervisory team - 1 Senior team - 1 SA team - 1	Provide L&O.	Supervise and conduct criminal investigations.	When directed, MP CS, L&O, and MWD support

Chapter 2

Battle Command

Command is the authority a commander in military service lawfully exercises over subordinates by virtue of rank and assignment. Leaders possessing command authority strive to use it with firmness, care, and skill.

FM 101-5-1

Battle command is the exercise of command in an operation against a hostile, thinking opponent. Battle command includes visualizing the current state and the desired end state, then formulating concepts of operations to get from one state to the other at the least cost. In addition to visualizing and formulating concepts, battle command encompasses assigning missions; prioritizing and allocating resources; selecting the critical time and place to act; and knowing how and when to make adjustments in the fight. Battle command enables MP commanders to lead, prioritize, and allocate assets required in support of the Army commander. MP commanders must observe, orient, decide, and act on their decisions quickly. Information is the key element in the battle-command process; therefore, the commander must have accurate and timely information upon which to base his decisions.

OVERVIEW

2-1. The battle command of MP units is typically decentralized due to the nature of their CS functions, METT-TC, and the needs of the Army commander. This places the burden of sound, timely decision making to the lowest levels. MP leaders must develop a keen sense of situational awareness and visualization, and they must constantly track the actions of supported units.

BATTLEFIELD VISUALIZATION

2-2. The ability to visualize the battlefield is a critical element of battle command. Battlefield visualization is an essential leadership attribute and is critical to accomplishing the mission. It is learned and attained through training, practice, experience, technical and tactical knowledge, and available battle-command technologies. It results when the MP commander understands the higher commander's intent, his assigned mission, the enemy, and the friendly force's capabilities and limitations. See Appendix D for further information on command technologies.

2-3. Battlefield visualization includes the MP commander's view of what his forces will do and the resources needed to do the mission. He envisions a sequence of actions that will cause his MP forces to perform at the desired end state. Ultimately, the MP commander's battlefield vision evolves into his intent and helps him develop his concept of operations.

COMMANDER'S INTENT

2-4. The commander's intent is a key part of Army orders. It is a clear, concise statement of what the force must do to succeed with respect to the enemy, the terrain, and the desired end state. It provides the link between the mission and the concept of operations by stating key tasks. These tasks, along with the mission, are the basis for subordinates to exercise initiative when unanticipated opportunities arise or when the original concept of operations no longer applies. MP leaders at all echelons must ensure that the mission and the commander's intent are understood two echelons down (see FM 101-5).

CONCEPT OF OPERATIONS

2-5. The commander's intent does not include the method by which the MP units will accomplish the mission. This method is called the commander's concept of operations. It must—

- Convey the commander's vision of how to accomplish the mission in a manner that allows his subordinates maximum initiative.
- Build around intelligence gathering and the precise employment of MP resources.
- Provide the basis for task organization, scheme of maneuver, terrain organization, tasks to subordinates, and synchronization.

COMMAND AND CONTROL RELATIONSHIPS

2-6. MP units are assigned to, attached to, or placed under the operational control (OPCON) of MP or other units they support. OPCON is the authority to perform command functions over subordinate forces. This includes organizing and employing commands and forces, assigning tasks, designing objectives, and giving authoritative direction necessary to accomplish the mission. MP C^2 relationships may be changed briefly to provide better support for a specific operation or to meet the needs of the supported commander. MP units may be placed under the OPCON of another unit commander for short-term operations. The MP unit remains in this relationship only as long as it is needed for that operation.

> MP support to the Bosnian municipal elections consisted of one division and two corps MP companies. These MP assets, attached to Task Force (TF) Eagle, were task-organized from different sources. The division MP company and the PM cell were organic to TF Eagle's mechanized infantry division headquarters, but the two corps MP units were from US Army Forces Command (FORSCOM) MP battalions in CONUS.

SUPPORT RELATIONSHIPS

2-7. MP units on the battlefield provide two types of support—general support (GS) and direct support (DS). Corps and EAC MP units provide GS to their respective corps/EAC subordinate commands. Light, airborne, and air-assault MP companies provide GS to their respective divisions. Heavy-division MP companies provide GS to the division rear and DS to the division's subordinate brigades.

STAFF RELATIONSHIPS

2-8. The PM for each level of command is that command's advisor on MP combat, CS, and CSS operations. The PM—

- Advises the commander and staff about MP abilities/capabilities.
- Supervises the preparation of plans and dictates policies.
- Coordinates MP operations.
- Assists and supervises the interaction of supporting and supported units.
- Reviews current MP operations.
- Coordinates with allied forces and HN military and civil police.
- Ensures that MP plans and operations supporting the commander's tactical plan are carried out.
- Recommends when and where to concentrate the command's MP assets.
- Supervises or monitors MP support in the command's AO.

2-9. The PM works daily with the commander and staff officers who employ MP resources and whose AORs influence MP support. The PM works closely with the coordinating staff at the appropriate command level to coordinate MP support. He ensures that MP planning is practical and flexible, that plans are coordinated with staff sections and subordinate commands, and that plans reflect manpower and resources needed by MP. (This includes the need for C^2, fire support, equipment, and supplies. It also includes construction, communication, transportation, and aviation support.) As new information is received, the PM reviews, updates, and modifies the plans. He ensures that the echelon commander gets the necessary MP support.

2-10. In the absence of specific directions or orders, the PM plans the use of MP assets. He evaluates the current operations and projects the future courses of action (COAs). He bases his plans on assumptions consistent with the commander's intent and a thorough knowledge of the situation and mission. The PM considers—

- METT-TC.
- Current estimates developed by the intelligence preparation of the battlefield (IPB) and the police information assessment process (PIAP).
- The environment within the AO. This includes the climate, the terrain, and obstacles. It also includes the legal authority and status of the force; the width, depth, size, and location of built-up areas; and the attitudes and abilities of the local populace.

- The types of units operating in the area (to include joint, combined, multinational, and interagency units) and the missions and capabilities of these units. This knowledge is imperative to understand their capability to counter threats in their area.
- The specific missions of MP units in the area and the impact that rear-area security operations will have on the ability of these units to perform other functions.
- Personnel, vehicles, and equipment in the MP units.

2-11. Coordination and communication between the PM and Army commanders is essential. Such actions ensure timely and efficient MP support to all levels of command during any operation. The informal, technical chain of coordination is an open line of communication between PMs at different echelons. The informal chain of coordination fosters cooperation and help among the MP elements at each echelon. For instance, when the division PM needs more assets to accomplish added missions, he initiates coordination with the corps PM. If the corps PM can provide support, the division PM formalizes his request for assistance through the division Assistant Chief of Staff, G3 (Operations and Plans) (G3).

Chapter 3

The Threat

In the 40-odd years of the Cold War, in many locations around the world, the Army performed a deterrent role as part of the containment strategy. In other places, at other times, the Army fulfilled the Nation's expectation in operations too small to be called "wars," although no less dangerous. To the soldier on the ground, Operations Urgent Fury in Grenada and Just Cause in Panama were indistinguishable from combat operations of their forefathers. Operations Provide Comfort in Iraq and Restore Hope in Somalia, although peace operations, also proved to be dangerous.

FM 100-1

The end of the Cold War has reduced, but not eliminated, the most immediate threat to the security of the US and other western nations. However, the absence of a dominant, identifiable threat has produced a far more complex and confusing strategic environment than the one that was present during the Cold War. Forward-deployed and CONUS-based ARFOR and civilians are and will continue to be engaged in a range of military actions. These actions stem from deterring conflicts to conducting peacetime engagement operations to providing support to civil agencies at home and abroad.

OVERVIEW

3-1. During the past decade, the US has deployed forces in multiple operations that have included crisis response in combat situations as well as participation in noncombat activities. The Army's presence in South Korea, Saudi Arabia, and Kuwait and its deployments to Somalia, Haiti, Bosnia, and Kosovo are clear indicators that the military must be prepared to face not only the traditional threat, but also a nontraditional, nonecheloned enemy. To support Army commanders successfully, MP leaders must understand the nature and complexity of these threats and how they can potentially affect the desired strategic, operational, and tactical end states.

REAR-AREA AND SUSTAINMENT OPERATIONS

3-2. The rear area for any particular command is the area extending forward from its rear boundary to the rear of the area assigned to the next lower level of command. This area is provided primarily for the performance of support functions. Operations in the rear area assure freedom of action and continuity of operations, sustainment, and C^2. Sustainment operations are those that enable shaping and decisive operations by assuring freedom of action and continuity of operations, CSS, and C^2 (see FM 3-0). Sustainment operations include the following elements:

- CSS.
- Rear-area and base security.
- Movement control.
- Terrain management.
- Infrastructure development.

3-3. During the Cold War, the danger to rear areas included forces that would be deployed in support of major soviet-style operations. The adversaries using the soviet model could be expected to engage in intense combat activity in their enemy's rear area. Their forces were prepared to penetrate into the enemy's rear and to attack and destroy its reserve forces and rear-area installations. To protect the rear areas, the MP were among the first mobile fighting forces available to the battlefield commander and thus, a source of combat power. Today, the Army commander uses the MP's flexibility and their modular-force training, adaptability, and mobility to serve as a combat multiplier throughout his entire AO. During sustainment operations, the MP perform all functions to ensure freedom of maneuver in support of the overall operational effort.

3-4. Failure to protect our forces during sustainment operations normally results in failure of the entire operation. Sustainment operations determine how fast ARFOR reconstitute and how far they can exploit success. The likelihood of MP units encountering the enemy and engaging in direct combat (not only in the rear area, but also during sustainment operations) cannot be underestimated.

3-5. Threats to rear-area and sustainment operations exist throughout the full spectrum of military operations. These threats may be related or independently engaged, but their effects are frequently cumulative. Threats to rear-area and sustainment operations are usually theater-dependent and are not limited to those outlined in this manual. Joint Publication (JP) 3-10 further discusses the threat in the rear area. Although JP 3-10 defines the threat in the context of a JRA, MP leaders can expect the same level of activity anywhere that US forces are deployed.

RECEPTION, STAGING, ONWARD MOVEMENT, AND INTEGRATION OPERATIONS

3-6. Reception, staging, onward movement, and integration (RSOI) operations consist of essential and interrelated processes in the AO that transform arriving personnel and materiel into forces capable of meeting operational requirements. During RSOI operations, the threat encountered will depend mostly on the type of entry, the nature of the operation, and the enemy. During major contingencies, forces deploy from power-projection platforms within the US or forward bases. The PM must plan MP support during the initial stages of the deployment to ensure the protection of follow-on forces and the detection of potential threats (see FM 100-17-3).

3-7. MP support to RSOI operations includes, but is not limited to—

- Conducting AS operations to counter or prevent enemy actions against marshalling and staging areas.
- Conducting convoy, airport, and rail security operations.
- Conducting populace- and resource-control operations.
- Conducting other physical-security and force-protection measures.
- Conducting other MP functions (as determined by the PM).

THREAT LEVELS

3-8. The threat is divided into three levels. These levels provide a general description and categorization of threat activities, identify the defense requirements to counter them, and establish a common reference for planning guidelines. MP leaders must understand that this does not imply that threat activities will occur in a specific sequence or that there is a necessary interrelationship between each level.

LEVEL I

3-9. Level I threats include the following types of individuals or activities:

- **Enemy-controlled agents.** Enemy-controlled agents are a potential threat throughout the rear area. Their primary missions include espionage, sabotage, subversion, and criminal activities. Their activities span the range of military operations and may increase during both war and military operations other than war (MOOTW). These activities may include assassinating or kidnapping key military or civilian personnel or guiding special-purpose individuals or teams to targets in the rear area.
- **Enemy sympathizers.** Civilians sympathetic to the enemy may become significant threats to US and multinational operations. They may be the most difficult to neutralize because they are normally not part of an established enemy-agent network, and their actions will be random and unpredictable. During war and MOOTW, indigenous groups sympathetic to the enemy or those simply opposed to the US can be expected to provide assistance, information, and shelter to guerrilla and enemy unconventional or special-purpose forces operating in the rear area.
- **Terrorism.** Terrorists are among the most difficult threats to neutralize and destroy. Their actions span the full spectrum of military operations.
- **Civil disturbances.** Civil disturbances, such as demonstrations and riots, may pose a direct or indirect threat to military operations. Although this threat may not be of great impact during war, it may significantly change and affect MOOTW.

LEVEL II

3-10. Level II threats include the following types of forces:

- **Guerilla forces.** Irregular and predominantly indigenous forces conducting guerrilla warfare can pose a serious threat to military forces and civilians. They can cause significant disruptions to the orderly conduct of the local government and services.

- **Unconventional forces.** Special-operations forces (SOF) are highly trained in unconventional-warfare techniques. They are normally inserted surreptitiously into the rear area before the onset of an armed conflict. They establish and activate espionage networks, collect intelligence, carry out specific sabotage missions, develop target lists, and conduct damage assessments of targets struck.
- **Small tactical units.** Specially organized reconnaissance elements are capable of conducting raids and ambushes in addition to their primary reconnaissance and intelligence-gathering missions. Small (size or capability), bypassed conventional units, as well as other potential threat forces, are also capable of conducting raids and ambushes to disrupt operations.

LEVEL III

3-11. Level III threats are made up of conventional forces. Potential threat forces are capable of projecting combat power rapidly by land, air, or sea deep into the rear area. Specific examples include airborne, heliborne, and amphibious operations; large, combined-arms, ground-forces operations; and bypassed units and infiltration operations involving large numbers of individuals or small groups infiltrated into the rear area, regrouped at predetermined times and locations, and committed against priority targets. Level III forces may use a combination of the following tactics as a precursor to a full-scale offensive operation:

- **Air or missile attack.** Threat forces may be capable of launching an air or missile attack throughout the rear area. It is often difficult to distinguish quickly between a limited or full-scale attack before impact; therefore, protective measures will normally be based on the maximum threat capability.
- **Nuclear, biological, and chemical (NBC) attack.** Commanders must be aware that NBC munitions may be used in conjunction with air, missile, or other conventional-force attacks. The NBC weapons could also be used at Level I or II by terrorists or unconventional forces in order to accomplish their political or military objectives.

THREAT-LEVEL MATRIX

3-12. Table 3-1 lists the threat levels and their likely appropriate responses. The threat levels listed are based on the type of threat. The table should not be construed as restricting the response options to any particular threat.

THREAT PRIORITIES

3-13. The threat will attempt to perform the following operations against targets in the rear area:

- Detect and identify targets.
- Destroy or neutralize operational weapons-system capabilities.
- Delay or disrupt the timely movement of forces and supplies.

Table 3-1. Threat Levels

Threat Level	Example	Response
I	Agents, saboteurs, sympathizers, and terrorists	Unit, base, and base-cluster self-defense measures
II	Small tactical units, unconventional-warfare forces, guerrillas, and bypassed enemy forces	Self-defense measures and response forces with supporting fires
III	Large tactical-force operations (including airborne, heliborne, amphibious, infiltration, and bypassed enemy forces)	Timely commitment of a TCF

- Weaken the friendly force's C^2 network.
- Disrupt support to combat forces.
- Set the stage for future enemy operations.
- Create panic and confusion throughout the rear area.

3-14. Typical examples of enemy priority targets include the following:

- NBC-weapons storage sites and delivery systems.
- Key command, control, communications, computers, and intelligence (C4I) facilities.
- Air-defense artillery (ADA) sites.
- Airfields and air bases.
- Port facilities.
- Main supply routes (MSRs) and MSR checkpoints.
- Key LOC.
- Reserve assembly areas (AAs).
- Troop barracks.
- Critical civilian and logistics facilities.

THREAT LOCATION

3-15. The fact that the Cold War has ended does not imply that our traditional threat has ended. North Korea and Iraq are constant reminders of this fact. For the near future, Army commanders will fight units with Cold-War-era equipment and tactics. The Army trains and is prepared to fight an enemy capable of interfering with our freedom of maneuver throughout the battlefield. On an extended battlefield with asymmetric threats, the danger to high-value assets (HVAs) (including CSS, C^2, communication nodes, and MSRs) only increases. The idea that the danger to the rear area decreases as you travel farther away from the forward edge of the battle area (FEBA) is not true. Threat intensity does not depend on geographical location; it depends on what operations the enemy believes must be initiated (and to what degree) to achieve its objective in the rear area. Military commanders depend on the MP to delay and defeat threats in their AO with a mobile reaction force.

COMMUNICATIONS ZONE

3-16. The nature of the COMMZ will encourage Level I and II threats to concentrate along the LOC and other areas of military significance. MP units will encounter an enemy that is capable of disrupting operations throughout the COMMZ while employing terrorist activities, enemy-controlled agent activities, enemy sympathizers, and saboteurs. If the enemy is Level III capable, MP leaders must expect infiltrations and air, missile, or NBC attacks as a precursor to a major Level III operation.

CORPS REAR AREA

3-17. The activities in Levels I and II will be similar in composition and density as in the COMMZ, but they will target key corps units, key facilities, and corps sustainment capabilities. The threat activities, especially at smaller unit levels, may even precede hostilities. MP leaders must be alert and prepared to encounter unconventional forces conducting diversionary or sabotage operations and small combat units conducting raids, ambushes, or reconnaissance operations or collecting special warfare intelligence. With the fast tempo of offensive operations, MP leaders must also be alert and prepared to encounter bypassed forces that can disrupt operations in the corps rear area.

DIVISION REAR AREA

3-18. The division rear area (DRA) contains many types of CS and CSS units and conducts many complex operations. As in the COMMZ and the corps rear area, the full spectrum of Level I, II, and III activities may occur in the DRA. The main target will be the division's HVA (including key C^2 facilities; airfields; artillery, aviation, and air-defense assets; LOC; and essential CSS units). The threat may conduct diversionary attacks, sabotages, raids, ambushes, and reconnaissance operations to affect the commander's freedom of maneuver and the continuity of operations. Unlike corps MP, the likelihood of division MP encountering bypassed enemy forces is expected. Failure to delay or defeat these forces will impact division operations.

OTHER TYPES OF THREATS

3-19. As US forces are deployed throughout the world, they will have to face nontraditional, asymmetric threats (other than those listed in Table 3-1, page 3-5) that may be geographically specific. As part of situational awareness, and in coordination with military intelligence (MI) and CID personnel, MP leaders must evaluate and assess the impact of these threats in their AO. A TO is vulnerable to any or a combination of the following threats:

- National or international organized crime.
- Narcotics traffickers.
- Narcotics terrorists.
- Extremist groups.
- Paramilitary groups.
- Ethnic or religious disputes.
- Trade in illegal weapons or strategic materials.

3-20. MP leaders must be aware that other threats exist and that they have the same potential as the Level I and II threats to disrupt operations in rear-area or sustainment operations. In some instances, the above threats' capabilities or the massing of personnel may have the same potential threat as a Level III threat.

COUNTERING THE THREAT

3-21. The disruption of rear-area and sustainment operations directly affects military efforts. Three types of forces may be used to counter the threat in these areas—a base/base-cluster self-defense force, a response force, or a tactical combat force (TCF).

BASE/BASE-CLUSTER SELF-DEFENSE FORCE

3-22. A base cluster is established when the appropriate echelon rear-operations cell or command post (CP) places geographically contiguous or noncontiguous bases under the control of a headquarters. The base cluster becomes the next higher tactical C^2 headquarters of those bases. The rear-operations cell or the rear CP may also establish a base cluster for a corps support group (CSG), an area support group (ASG), or other CSS units operating in the corps or division rear areas.

3-23. US ARFOR have the inherent responsibility to contribute as many forces as possible for base defense and local security for themselves and their facilities, installations, and activities. Each base and base-cluster commander must develop a defense plan to detect, defeat, and minimize the effects of Level I and limited Level II threat attacks on his base or base cluster (including NBC attacks). To maximize the unit's mission accomplishment, defense plans must be flexible and allow for differing degrees of security based on the probability of threat activity. Defense plans are given to MP units operating near the base or base cluster. The base commander most often employs a series of defense measures providing internal and perimeter security. His internal reaction forces use organic weapons to neutralize and defeat most low-level threat activity. Although not fully equipped to engage major conventional or unconventional enemy forces that may confront him, a base commander must deploy his personnel to defend themselves until MP, HN, local police, or combat forces (if available) can respond.

RESPONSE FORCE

3-24. A response force is summoned when the base or base cluster is faced with threat forces that are beyond their self-defense capability. If the MP are the designated response force, they must—

- Coordinate with the supported bases or base-cluster commanders to conduct a joint IPB.
- Review base and base-cluster self-defense plans.
- Exchange signal-operating-instructions (SOI) information.
- Identify MP contingency plans to counter likely enemy activities.
- Integrate ADA, engineer, chemical, field-artillery (FA), Army-aviation, and close-air-support (CAS) fire support into their plans (if available).

3-25. MP units help the base or base cluster return to its primary mission by defeating Level II threats. MP units closely watch likely avenues of enemy approach, possible landing zones (LZs), drop zones (DZs), C^2 facilities, and other key installations. They accomplish this through the MMS, AS, and PIO functions.

3-26. If MP units are not the designated response force, they may become the initial response force for units within their AO. When this occurs, they block, delay or, if possible, destroy enemy elements within their capability. If the attack is by a larger or more capable force, they will maintain contact and continue to develop the situation or delay until the appropriate response force appears or the battlefield commander commits the TCF.

TACTICAL COMBAT FORCE

3-27. When the MP response force encounters or engages threats beyond its ability to defeat, it immediately notifies the higher headquarters. The battlefield commander will then evaluate the situation and commit the TCF to defeat the Level III threat. The TCF is normally a combined-arms organization tailored by the corps or division G3, based on METT-TC. The TCF normally receives fire, aviation, or other support needed to fight and defeat the threat. Once the TCF is identified and before it is committed to battle, it will conduct direct coordination with the MP or other response forces regarding the exchange of reconnaissance information, battle-handoff procedures, and contingency plans for TCF operations. Once the TCF is committed, the MP unit performing as the response force becomes OPCON to the TCF commander.

MP AS THE TCF

3-28. The MP brigade or battalion is capable of conducting TCF missions only when properly augmented. MP augmentation may be in the form of fires, small combat units, aviation assets, or CAS. The specific type of augmentation is METT-TC dependent. Additionally, the MP C^2 headquarters must receive the respective liaison officers to ensure that augmentation forces are synchronized and employed according to their capabilities. The MP commander's situational awareness and battlefield visualization are key elements to TCF operations. Once designated as the TCF, the MP unit commander establishes liaison with the appropriate rear CP to obtain—

- The current rear-area IPB.
- The friendly unit disposition.
- Defense plans.
- Priorities for protection.
- The fire-support plan.

3-29. Based on the above information, the MP commander conducts his own IPB and develops a concept of operations. He then forwards it to the appropriate higher echelon for coordination and approval.

3-30. The MP's ability to employ organic MP assets as part of the TCF is limited by the following factors:

- **MP availability.** Normally, all MP assets available are committed at all times. The specific function and scope that MP units perform during the operation are determined by the Army commander's needs, the intensity of the conflict, and the availability of MP resources. The commander, with advice from the MP leader, must decide which MP operations must be scaled back, delayed, or shifted before the MP unit can be designated as part of the TCF.
- **MP dispersion.** MP units are normally displaced over a large geographical area. Technological capabilities and mobility allow them to operate over great distances. In today's battlefield, a typical MP company employment covers between 1,000 and 1,200 square kilometers and performs numerous missions in support of all five functions. The distance between elements, the reprioritization and movement of other MP units, the difficult terrain, poor roads, and bad weather may slow down the MP's commitment as a TCF.

Chapter 4

Military Police Functions

Military police support the Army commander's mission to win the battle. They help the commander shape the battlefield so that he can conduct decisive operations to destroy enemy forces, large or small, wherever and whenever the Army is sent to war.

MAJ(P) Anthony Cruz

The MP Corps supports shaping and sustainment operations while performing its five basic functions as a flexible, economy-of-force organization. Through these functions, MP units are able to provide the commander with an array of CS operations across the full spectrum of military operations (see Table 4-1).

Table 4-1. MP Functions

	Functions				
	MMS	**AS**	**PIO**	**L&O**	**I/R**
Subtasks	Support to river-crossing and breaching operations and passage of lines Straggler and dislocated-civilian control Route R&S MSR regulation enforcement	Recon operations ADC Base/air-base defense Response-force/ TCF operations Critical site, asset, and HRP security Force protection/ physical security Antiterrorism	IPB support PIAP Active and passive roles Information collection and dissemination Joint, interagency, and multinational coordination	Law enforcement Criminal investigations US Customs operations Related L&O training	EPW/CI handling US military prisoner handling Populace and resource control Dislocated civilians
NOTE: Subtasks not all-inclusive.					

OVERVIEW

4-1. MP assets are limited. Specific functions are performed at any given time and are determined by the supported commander's need, the intensity of the conflict, and the availability of MP resources. The supported commander, through the command's PM, sets the priorities for MP operations.

4-2. The PM (based on METT-TC and the available assets) continuously evaluates the trade-off between the MP support that the commander requires and the MP support that can be provided. To meet the priorities set by the commander's tactical plan, the PM recommends the allocation and employment of MP assets for MP combat, CS, and CSS operations.

MANEUVER AND MOBILITY SUPPORT

4-3. The MMS function involves numerous measures and actions necessary to support the commander's freedom of movement in his AOR. The MP expedite the forward and lateral movement of combat resources and ensure that commanders get forces, supplies, and equipment when and where they are needed. This is particularly important in the modern battlefield where there is a greater geographical dispersal of forces and lengthened LOC.

4-4. The MP maintain the security and viability of the strategic and tactical LOC to ensure that the commander can deploy and employ his forces. The MP support the commander and help expedite military traffic by operating traffic-control posts (TCPs), defilades, or mobile patrols; erecting route signs on MSRs or alternate supply routes (ASRs); or conducting a reconnaissance for bypassed or additional routes. The MP move all units quickly and smoothly with the least amount of interference possible.

4-5. As part of the MMS function, the MP support river-crossing operations, breaching operations, and a passage of lines. They also provide straggler control, dislocated-civilian control, route reconnaissance and surveillance (R&S), and MSR regulation enforcement.

RIVER-CROSSING OPERATIONS

4-6. US forces conduct river-crossing operations to move a large force across a river obstacle with a minimum loss of momentum. The MP play a vital role by assisting the force commander in crossing the river as quickly and efficiently as possible. The crossing is usually planned and conducted by the headquarters directing the crossing. As such, a division crossing operation is conducted by a corps. Whether a brigade or division is crossing, the division MP company may also cross to provide uninterrupted support to the division. In these instances, there is a total reliance on the corps MP to support the crossing. The same is true for breaching operations and a passage of lines.

4-7. MP support for river-crossing operations reduces the crossing time and promotes the efficient movement of vehicles. It reduces congestion, speeds the crossing, and enables the maneuver commander to continue his momentum toward his primary objective. The MP establishes staging and holding areas and TCPs to control movement to and from these areas (according to the traffic-control plan). The MP may be called on to provide security for crossing forces at the crossing sites. In most cases, the MP TCPs and engineer regulation points (ERPs) are located on both sides of the river to improve communications and coordination between the units.

4-8. MP employment for river crossing is influenced by METT-TC. The number and placement of MP assets supporting a river-crossing operation varies with the size of the crossing force, the direction of the crossing (forward or retrograde), and the degree of enemy resistance expected or encountered.

4-9. The MP operating inside the crossing areas are OPCON to the crossing-area commander for the duration of the operation. The MP operating outside of the crossing area are under the command of their appropriate echelon commander.

4-10. The main thrust of MP support to river-crossing operations is within the immediate river-crossing site. The MP direct units to their proper locations (such as holding areas and staging areas) and ensure that units move through the area within the time listed on the movement schedule. This is a highly critical aspect of river crossing because the number of crossing sites is limited. MP assets are placed where they can stress MMS operations on MSRs leading into the crossing area.

4-11. The MP also provide AS to allow crossing forces to cross the river without losing momentum or forces. On both near and far sides, the MP are used to recon the crossing unit's flanks and rear to enhance security (see FM 19-4).

BREACHING OPERATIONS

4-12. Breaching operations are conducted to allow forces to maneuver through obstacles. Obstacle breaching is the employment of a combination of tactics and techniques to advance an attacking force to the farside of an obstacle that may be covered by fire. It is perhaps the most difficult combat task a force can encounter. Breaching operations begin when friendly forces detect an obstacle, and they end when the battle handover has occurred between the follow-on forces and a unit conducting the breaching operation (see FM 90-13-1).

4-13. The MP support breaching operations in numerous ways. MP assets are employed based on METT-TC, the available resources, and the commander's priorities. As a minimum, MP support may include, but is not limited to—

- Establishing TCPs along routes leading to or departing from the breaching site.
- Establishing holding areas.
- Establishing TCPs at the breaching site.
- Assisting engineers with temporary route signs.
- Establishing straggler-control operations.
- Conducting AS operations.

4-14. The most critical MP support is provided at the breaching site. The MP provide the commander with a means to control traffic flow to appropriate lanes. When multiple lanes branch off of a single far-recognition marker, the MP assist in directing the formation through various lanes. They also assist in modifying the traffic flow when lanes have been closed for maintenance or expansion. The MP conduct close coordination with the crossing-force commander and the TF commander executing the breaching operation. The MP enable the commander to make last-minute changes in traffic flow, thereby giving him increased flexibility to react to the enemy situation.

PASSAGE OF LINES

4-15. A passage of lines is an operation in which a force moves forward or rearward through another force's combat positions with the intention of moving into or out of contact with the enemy. The passage of lines is a high-risk military operation that requires close coordination between the passing unit, the stationary unit, and the MP providing the support.

4-16. The MP help reduce confusion and congestion during a passage of lines. They provide security in areas surrounding passage points and passage lanes to ensure that the passing unit has priority for using routes to and through the areas. The headquarters directing the operations sets the route's priority. The MP can support a forward, rearward, or lateral passage of lines. Before the actual operation, the MP in the AO conduct an area or zone reconnaissance to become familiar with the routes to, through, and beyond the area of passage. This enables the MP to extend the commander's C^2 by providing directions at passage points and by guiding the units through the passing lanes. Maintaining unit integrity and reducing incidents of stragglers is vital to maintaining the passing unit's momentum in a forward passage of lines. The MP perform aggressive straggler- and dislocated-civilian-control operations to prevent possible infiltration of the enemy.

4-17. A passage of lines is usually planned and coordinated by the headquarters directing the passage. A division's passage of lines is planned and coordinated by the corps headquarters. The detailed plans are made and coordinated between the divisions involved. Close coordination between the division and corps PMs is essential. An MP unit may be the unit involved in passing through the lines of another unit. When conducting a delay of a Level II threat, the MP are likely to conduct a passage of lines with the TCF. To avoid fratricide, close coordination between the MP response-force commander and the TCF is imperative (see FM 19-4).

STRAGGLER CONTROL

4-18. Mobile patrols, TCPs, and checkpoint teams return stragglers to military control as part of their operations. Most stragglers are simply persons who become separated from their command by events in the CZ or while moving through the COMMZ. If a straggler is ill, wounded, or in shock, an MP must give him first aid and, if needed, call for medical evacuation (MEDEVAC). If a straggler is uninjured, an MP directs him to his parent unit or to a replacement unit (as command policies dictate). The MP ensure that stragglers attempting to avoid return to their units are escorted back to their command (as a minimum).

4-19. The MP set up special posts for straggler control following NBC attacks or major enemy breakthroughs that result in large numbers of lost, dazed, and confused military personnel. Mobile MP teams operate between posts, and they also direct or collect stragglers. Straggler collection points may be needed if many stragglers are present in a combat theater. If allied forces are present in the theater, each nation establishes a collection point for its own personnel. MP teams are aware of each allied location and are prepared to assist allied soldiers in returning to their respective command. The MP use available transportation assets to transfer stragglers from TCPs and checkpoints to a straggler collection point. At the collection point, they are screened and sorted for removal to a medical facility or returned to their units to reconstitute the tactical commander's combat force.

4-20. The MP report information about stragglers with whom they come in contact. This information is compiled by the MP headquarters and forwarded through appropriate channels to the higher command. Information given by stragglers that is of immediate tactical value is reported without delay.

DISLOCATED-CIVILIAN CONTROL

4-21. The MP expediting traffic on MSRs may encounter dislocated civilians that could hinder military traffic. The MP assist and divert dislocated civilians from MSRs and other areas to I/R facilities. They may also deny the movement of civilians whose location, direction of movement, or actions may be a threat to themselves or to the combat operation. The HN government is responsible for identifying routes for the safe movement of dislocated civilians out of an AO. If needed, the MP assist the civil-affairs unit and the HN in redirecting dislocated civilians to alternate routes established by the HN government.

4-22. The US forces do not assume control of dislocated civilians unless requested to do so by the HN or unless operating in an environment with a hostile government. When the senior US commander assumes responsibility, the MP coordinate with civil affairs to set up TCPs at critical points along the route to direct dislocated civilians to secondary roadways and areas not used by military forces. As required, MWD teams may be used as a show of force or as a deterrent to assist with uncooperative personnel.

ROUTE R&S

4-23. The MP conduct hasty and deliberate route reconnaissances to obtain information on a route and nearby terrain from which the enemy can influence troop movement. A route reconnaissance focuses on continually monitoring the condition of MSRs, ASRs, and other areas. MP patrols look for restricting terrain, effects of weather on the route, damage to the route, NBC contamination, and enemy presence or absence. When enemy activity is spotted, the MP report it, maintain surveillance, and develop the situation. To gather information for proposed traffic plans, they look at the type and number of available routes; and they check load classifications, route widths, obstructions, and restrictions.

MSR REGULATION ENFORCEMENT

4-24. The MP undertake MSR regulation enforcement to keep the routes free for DSS operations. MP units support the command's MSR regulation measures as stated in the traffic-regulation plan (TRP). The TRP contains specific measures to ensure the smooth and efficient use of the road network. It assigns military route numbers and names, the direction of travel, highway regulation points, and preplanned MP TCPs. Most importantly to the MP, it gives the route's control classification. The MP ensure that classified routes are used only by authorized traffic. Vehicles traveling on roads too narrow for their passage or on roads unable to support their weight can obstruct the route.

4-25. To expedite traffic on MSRs, the MP operate special circulation control measures such as—

- Temporary route signing.
- Static posts such as TCPs, roadblocks, checkpoints, holding areas, and defilades at critical points.
- Mobile teams patrolling between static posts and monitoring traffic and road conditions.

4-26. They also gather information on friendly and enemy activities and help stranded vehicles and crews. The MP also place temporary route signs to warn of hazards or to guide drivers unfamiliar with the route. Using these measures, the MP exercise jurisdiction over the road network in the AO and coordinate with the HN (whenever possible) to expedite movement on MSRs.

AREA SECURITY

4-27. The MP perform the AS function to protect the force and to enhance the freedom of units to conduct their assigned missions. The MP who provide AS play a key role in supporting forces in rear-area and sustainment operations. The MP are a response force that delays and defeats enemy attempts to disrupt or demoralize military operations in the AO. The MP's mobility makes it possible for them to detect the threat as they aggressively patrol the AO, MSRs, key terrain, and critical assets. The MP's organic communications enable them to advise the appropriate headquarters, bases, base clusters, and moving units of impending enemy activities. With organic firepower, the MP are capable of engaging in decisive operations against a Level II threat and delaying (shaping) a Level III threat until commitment of the TCF.

4-28. Throughout all aspects of the AS function, the MP perform counteractions to protect the force and to prevent or defeat a Level II threat operating within the MP's AO. MP countermeasures may include implementing vulnerability assessments, developing procedures to detect terrorist actions before they occur, hardening likely targets, and conducting offensive operations to destroy the enemy. The MP use checkpoints and roadblocks to control the movement of vehicles, personnel, and materiel and to prevent illegal actions that may aid the enemy. The use of these control measures serves as a deterrence to terrorist activities, saboteurs, and other threats. However, at the same time, checkpoints and roadblocks expose the MP to these potential threats. To counter this fact, the MP may upgrade or harden vehicles and defensive positions.

4-29. The MP provide combat power to protect the C^2 headquarters, equipment, and services essential for mission success. The MP provide the battlefield commander with a light, mobile fighting force that can move, shoot, and communicate against any threat. Major subtasks associated with the AS function include reconnaissance operations; area damage control (ADC); base/air-base defense; response-force operations; and critical site, asset, and high-risk personnel (HRP) security.

RECONNAISSANCE OPERATIONS

4-30. As part of their AS mission, the MP serve as the eyes and ears of the battlefield commander by seeking out the enemy and reporting information obtained by recon patrols. The MP conduct area and zone reconnaissances, screening, surveillance, and countersurveillance to gain information to help guard against unexpected enemy attacks in the AO. The MP monitor likely avenues of approach and potential LZs and DZs. They become familiar with towns and other populated areas, ridgelines, woods, and other terrain features from which the enemy can influence movements along road networks. The MP pay close attention to areas near facilities designated critical by the commander. These areas include key MSR bridges and tunnels, depots, terminals, logistics-support bases, ammunition supply points (ASPs), communications centers/nodes, and C^2 headquarters. The MWD teams provide explosive detection and personnel detection/tracking capabilities that enhance reconnaissance operations (especially in urban terrain).

AREA DAMAGE CONTROL

4-31. MP units take measures to support ADC before, during, and after hostile actions or natural and man-made disasters. The ADC actions integrate CS and CSS functions for many units. Engineers, medical personnel, and Army aviators work closely to ensure quick relief operations. The MP provide MSR regulation enforcement, refugee control, and some local security when required. As with reconnaissance operations, the MP may use MWD explosive-and personnel-detection capabilities to augment all MP missions in rear-area and sustainment operations.

BASE/AIR-BASE DEFENSE

4-32. The MP are the base and base-cluster commanders' links for detection, early warning, and employment against enemy attacks. The information gathered is dispersed throughout the rear area to help apprise the commander of enemy activities near bases. Base defense is the cornerstone of rear-area security. When the threat exceeds the base/base-cluster capability, the base/ base-cluster commander requests MP assistance through the appropriate C^2 element.

4-33. Air-base defense requires special MP coordination with the US Air Force (USAF). The MP treat air bases like any other base or base cluster. A USAF air base may house the base-cluster commander, or it may be a cluster by itself. The MP are responsible for the air base's external defense. Its internal defense is primarily the responsibility of the Air Force's security forces. The security force provides in-depth defense for weapons, weapons systems, command centers, personnel, and other priority resources established by the base commander.

4-34. The security force is trained and equipped to detect, delay, and deny Level I and II threats. If a Level III threat is present, the security force is tasked with delaying actions; however, the HN, a sister service, or other support must be employed to defeat this threat. If the security force requires assistance to defeat a Level II threat, it may rely on MP response forces or another response force to assist in the defense. If available, the MP response force will react to the air-base defense, just as it would for any other base or base cluster within the MP's AO. However, the key to successful MP employment depends on the critical exchange of information before and during the MP employment. Good communications, an understanding of the defense plan, and liaison operations are vital in preventing responding forces from entering a situation that could result in fratricide.

RESPONSE-FORCE OPERATIONS

4-35. The MP are the base and base-cluster commanders' response force against enemy attacks in rear-area or sustainment operations. The MP gather information about the enemy while performing missions throughout the AO. This information provides commanders with enemy activity near bases. When needed, the MP provide a mobile response force to respond to bases under attack and to destroy the enemy. A base commander's defense plan is the cornerstone for protecting rear-area and sustainment operations. The base commander is responsible for defeating all Level I threats. When this threat exceeds his capabilities, he requests MP support. The MP located near bases or patrolling or conducting AS operations consolidate their forces, respond as quickly as possible, and conduct combat operations to destroy the enemy. If needed, the MP conduct a battle handover to the TCF.

4-36. MP forces performing as a response force are capable of conducting the following offensive operations:

- A movement to contact.
- A hasty ambush.
- A hasty attack.
- A delay.
- A call for fire.
- A repel attack against critical sites.
- A defense of critical sites.

4-37. To conduct these missions, the MP consolidate into squads or platoons to delay, defeat, or defend against the threat. See FMs 71-3 and 71-100 for more information on battle-handover operations.

CRITICAL SITE, ASSET, AND HRP SECURITY

4-38. The MP perform their AS function across the entire designated AO. When the MP provide security around a critical site or asset, they usually provide a mobile security screen, taking advantage of its weapons and communications platforms. This standoff protection detects and defeats the threat before it can move within direct-fire range of the facilities. The MP may be tasked to provide detail security to key facilities, assets, and personnel.

4-39. The MP provide security to major CPs and other facilities within the AO. Their employment maximizes mobility, lethality, and communications capabilities as a security screen. They may be required to establish local AS measures (such as checkpoints and listening/observation posts) to further protect these facilities. The MP provide internal access-control points to critical facilities, and they act as a response force. When the critical CP relocates, the MP provide in-transit security. Other types of critical site security include ASPs; deep-water ports; petroleum, oil, and lubricants (POL) terminals and pipelines; trains and railways; and air bases.

4-40. The MP may provide convoy security for top-priority units transporting especially critical supplies to combat forces. MP assets should be employed primarily on aggressive patrolling, route, area, and zone reconnaissance measures that would create a safe and secure environment for all types of vehicular and unit movement.

4-41. The MP and the CID provide protective services to designated key personnel by providing access control to restricted areas within CPs, providing in-transit security, or providing static security measures around the clock. The MP coordinate with the CID when close-in protection of key personnel is needed. The MP and the CID also provide training for personal-protection countermeasures. The MWD teams may be employed to enhance MP and CID detection and protection capabilities.

INTERNMENT AND RESETTLEMENT

4-42. The Army is the Department of Defense's (DOD's) executive agent for all EPW/CI operations. Additionally, the Army is DOD's executive agent for long-term confinement of US military prisoners. Within the Army and through the combatant commander, the MP are tasked with coordinating shelter, protection, accountability, and sustainment for EPWs/CIs. The I/R function addresses MP roles when dealing with EPWs/CIs, dislocated civilians, and US military prisoners.

4-43. The I/R function is of humane as well as tactical importance. In any conflict involving US forces, safe and humane treatment of EPWs/CIs is required by international law. Military actions on the modern battlefield will result in many EPWs/CIs. Entire units of enemy forces, separated and disorganized by the shock of intensive combat, may be captured. This can place a tremendous challenge on tactical forces and can significantly reduce the capturing unit's combat effectiveness. The MP support the battlefield commander by relieving him of the problem of handling EPWs/CIs with combat forces. The MP perform their I/R function of collecting, evacuating, and securing EPWs throughout the AO. In this process, the MP coordinate with MI to collect information that may be used in current or future operations.

4-44. Although the CS MP unit initially handles EPWs/CIs, modular MP (I/R) battalions with assigned MP guard companies and supporting MWD teams are equipped and trained to handle this mission for the long term. A properly configured modular MP (I/R) battalion can support, safeguard, account for, guard, and provide humane treatment for up to 4,000 EPWs/CIs; 8,000 dislocated civilians; or 1,500 US military prisoners.

EPW/CI HANDLING

4-45. The MP are tasked with collecting EPWs/CIs from combat units as far forward as possible. The MP operate collection points and holding areas to temporarily secure EPWs/CIs until they can be evacuated to the next higher echelon's holding area. The MP escort-guard company assigned to the MP brigade (I/R) evacuate the EPWs/CIs from the corps's holding area to the COMMZ's internment facilities. The MP safeguard and maintain accountability, protect, and provide humane treatment for all personnel under their care.

4-46. In a mature TO, MP (I/R) units process EPWs/CIs collected by MP teams and other units in the CZ. MP guard companies assigned to the MP (I/R) units guard EPWs/CIs at designated camps (see FM 19-40).

POPULACE AND RESOURCE CONTROL

4-47. Populace and resource control (PRC) denies adversaries or insurgents access to the general population and resources. The MP supports civil-affairs personnel and the tactical commander in planning and conducting PRC programs employed during all military operations. These programs may consist of curfews, movement restrictions, resettling dislocated civilians, licensing, ration control, regulation enforcement, amnesty programs, inspecting facilities, and guarding humanitarian-assistance distributions. The MP also direct dislocated civilians to resettlement camps where they are cared for while NGOs work to coordinate their relocation.

4-48. The MP's security capability, acceptability, and interface with the populace make them suitable as the primary forces of choice in these operations. The MP I/R units are specifically designed to fill this need (see FM 41-10).

US MILITARY PRISONERS CONFINEMENT

4-49. The MP detain, sustain, protect, and evacuate US military prisoners. Whenever possible, soldiers awaiting trial remain in their units, unless reasonable grounds exist to believe that they will not appear at the trial, the pretrial hearing, or the investigation or that they will engage in serious criminal misconduct. Under either of these two pretrial confinement instances, the commander must also reasonably believe that a less severe form of restraint (such as conditions of liberty, restriction in lieu of apprehension, or apprehension) are inadequate. When these circumstances exist and other legal requirements are met, US military personnel may be placed in pretrial confinement under the MP's direct control. Convicted military prisoners are moved as soon as possible to confinement facilities outside of the AO.

4-50. MP confinement operations parallel (but are separate from) the MP's EPW/CI operations. No member of the US armed forces may be placed in confinement in immediate association with an EPW or other foreign nationals who are not members of the US armed forces. A confinement facility is maintained within the TO only if distance or the lack of transportation requires such a facility. When military prisoners are retained in theater, temporary field detention facilities may be established in the CZ and a field confinement facility in the COMMZ (see FM 19-40 and AR 190-47).

LAW AND ORDER

4-51. The L&O function consists of those measures necessary to enforce laws, directives, and punitive regulations. The MP's L&O function extends the battlefield commander's C^2. The MP, in close coordination with the CID, work to suppress the chance for criminal behavior throughout the AO. By coordinating and maintaining liaison with other DOD, HN, joint, and multinational agencies, the MP at all levels coordinate actions to remove conditions that may promote crime or that have the potential to affect the combat force. Crime-prevention measures and selective enforcement measures are also performed as part of other functions. For example, the MP investigate traffic accidents and regulate traffic as part of the MMS function. The L&O function includes major areas such as law enforcement, criminal investigations, and support to US Customs operations. The primary units conducting L&O are the L&O detachments, customs teams, and CID units. Both the MWD team and the MP company (CS) also support the L&O function.

LAW ENFORCEMENT

4-52. Law-enforcement operations assist the battlefield commander in preserving his force. The MP dedicate assets to conduct law-enforcement operations based on the commander's needs. Since MP L&O and CID assets may be limited during the initial stages of any operation, the PM depends on the supported commander's development of an effective crime-prevention program and uses established investigative tools (such as inquiries and AR 15-6 investigations) to enforce rules. This will allow the PM to employ limited MP assets to perform other functions. Law-enforcement operations include responding to civil disturbances, conducting raids, investigating traffic accidents, conducting vehicle searches, supporting the commander's force-protection program, and providing support to HN and civil-enforcement agencies. Law enforcement also includes employing special-reaction teams (SRTs), marksman/observer (M/O) teams, and MWD teams.

4-53. The Army conducts counterdrug-support operations that generally fall within several DOD counterdrug-mission categories. The MP support the Army's role rather than directly participating in civil law-enforcement activities (such as searches, seizures, and arrests). When tasked, the MP provide training to law-enforcement agencies in common soldier skills, physical security, and tactical planning and operations. US Code (USC) Title 18, Posse Comitatus Act, Section 1385 prohibits the use of DOD assets to enforce civilian law (federal and state) except in cases and under circumstances expressly authorized by the Constitution or by an act of Congress.

4-54. In multinational operations, the MP may assist with the creation of multinational police units. Circumstances that may support the establishment of these police forces include existing or negotiated terms of international agreements or security-assistance programs, a multinational operational agreement, or appropriate military directives. The MP provide the capability to train foreign MP and/or reconstitute indigenous constabulary forces as part of stability and support operations. The MP can provide the initial mentoring to these forces and provide temporary, emergency law-enforcement capabilities until the foreign military or civilian police forces are functional (see FM 100-8).

CRIMINAL INVESTIGATIONS

4-55. The MP investigate offenses against US forces or property committed by persons subject to military law. Investigations against minor crimes (such as low-value, personal-property thefts or simple assaults) are normally investigated by the MP's L&O detachment. Investigations against major incidents involving wrongful death, serious bodily injury, and war crimes are referred to special agents of the USACIDC. The USACIDC conducts death investigations in the absence of HN agreements or in conjunction with the HN. The USACIDC special agents are authorized to investigate any alleged criminal conduct in which there is an Army interest unless prohibited by law or higher authority.

4-56. The USACIDC's investigative authority and investigative responsibility outside of the US are determined by international treaty or agreement (including status of forces agreements [SOFAs]), the policies of the HN government, the US ambassador, and AR 195-2. In the absence of such provisions, the following guidelines apply:

- On Army-controlled installations, the USACIDC has the authority to investigate alleged crimes.
- Outside of an Army-controlled installation, the USACIDC may investigate after coordinating with HN authorities.

4-57. In all environments, the USACIDC has the responsibility to investigate all felony crimes involving Army personnel, DA civilians and agencies, and companies working for the Army. The USACIDC investigates war crimes and crimes involving personal and government property affecting the Army's mission (logistics security [LOGSEC]). Other investigations (such as those based on international treaties, SOFAs, and joint investigations with the HN) may be undertaken if requested by the supported commander in support of the overall Army mission. See Chapter 9 for a complete discussion of the USACIDC.

US CUSTOMS OPERATIONS

4-58. The MP support the US Customs Service (USCS), the US Department of Agriculture (USDA), other federal agencies, joint staffs, and commanders who enforce the laws and regulations of the US concerning customs, agriculture, and immigration border clearances. Support to the USCS also includes assistance to federal agencies to eliminate the illegal traffic of controlled substances and other contraband through Army channels. MP support to customs operations are normally performed by specially trained MP customs teams. Although other MP units are not trained in all facets of customs operations, they may assist MP customs teams, the USCS or the USDA, and other federal agencies in the enforcement of applicable laws and regulations. When tasked, the MP/CID supports the investigation of violations of US Codes, DOD or DA regulations, and applicable provisions of SOFAs.

4-59. The MP report violations of customs laws, regulations, inspections, and investigative results to the installation's PM, the supported commander, and affected units. During redeployment from outside the continental US (OCONUS) to CONUS installations, the MP support the USCS or USDA efforts to ensure that personnel, equipment, and materiel meet customs, immigration, and agriculture requirements as stated by all applicable laws and regulations. As with other functions, MWD teams may be employed in support of customs operations for the detection of explosives or narcotics.

> During operation Just Cause, an MP platoon temporarily assumed the customs mission at the main terminal of the Torrijos International Airport located just outside of Panama City. Their mission supported the air evacuation of foreign-national civilians and the redeployment of some of the initial-entry US forces.

POLICE INTELLIGENCE OPERATIONS

4-60. The PIO function supports, enhances, and contributes to the commander's protection program, situational awareness, and battlefield visualization by portraying relevant threat information that may affect his operational and tactical environments. This threat information—whether it is police, criminal, or combat information—is gathered while conducting MP functions. The PIO function—

- Demonstrates the MP's/CID's capability to collect relevant threat information actively or passively.
- Ensures that all information collected while conducting MMS, AS, I/R, and L&O functions continues to be reported through the proper channels so that it can be analyzed by the Intelligence Officer (US Army) (S2) or the Assistant Chief of Staff, G2 (Intelligence) (G2) with support from the appropriate MP echelon.
- Coordinates with USACIDC elements to employ data developed by the USACIDC's programs. These programs include—
  - The Combating Terrorism Program as outlined in AR 525-13 and CIDR 195-1.
  - The Criminal Intelligence Program (CIP).
  - Personal-security vulnerability assessments (PSVAs).
  - A crime threat analysis.
  - Logistics-security threat assessments (LSTAs).

NOTE: The MP/CID must ensure that criminal information is released according to existing controls and restraints.

- Maintains constant liaison and communication with the higher echelon S2/G2; psychological-operations (PSYOP) units; HN police and other law-enforcement agencies; joint, combined, interagency, and multinational forces; the staff judge advocate (SJA); the CMOC; civil-affairs teams; and the force-protection officer.

4-61. The MP brigade commander, the battalion commander, and the PM are responsible for the PIO function. As such, each one must determine the best way to employ the available staff resources to monitor the execution of the PIO function within his command.

IPB PROCESS AND PIO FUNCTION

4-62. The IPB is a systematic, continuous process for analyzing the threat and environment in a specific geographic area. It is designed to support staff estimates and military decision making. Applying the IPB process helps the commander selectively apply and maximize his combat power at critical points in time and space on the battlefield by—

- Determining the threat's likely COA.
- Describing the environment the unit is operating within and the environmental effects on the unit.

4-63. The IPB process consists of the following four steps:

- **Define the battlefield environment.** The S2/G2 identifies the battlefield characteristics that will influence friendly and threat operations, establishes the limits of the area of interest (AOI), and identifies gaps in current intelligence holdings.
- **Describe the battlefield effects.** The S2/G2 evaluates the environmental effects with which both sides must contend. The S2/G2 identifies the limitations and opportunities that the environment offers on the potential operations of friendly and threat forces. This evaluation process focuses on the general capabilities of each force until COAs are developed later in the IPB process. This environmental assessment always includes an examination of terrain and weather, but it may also include discussions of characteristics of geography and infrastructure and their effects on friendly and threat operations.
- **Evaluate the threat.** The S2/G2 and his staff analyze the command's intelligence holdings to determine how the threat normally organizes for combat and conducts operations under similar circumstances. When facing a well-known threat, the S2/G2 can rely on historical databases and threat models. When operating against a new or less known threat, he may need to develop his intelligence databases and threat models concurrently.
- **Determine the threat's COA.** Given what the threat normally prefers to do and the effects of the specific environment in which he is operating, his likely objectives and the COAs available to him are determined. The S2/G2 develops enemy COA models that depict the threat's available COAs. He also prepares event templates and matrices that focus intelligence and identify which COA the enemy will execute (see FM 34-130).

4-64. Although the S2/G2 has the staff responsibility for the command's IPB, he is not the only one who conducts or needs to understand and use the IPB. Every Army commander and staff member must understand and apply the IPB process during the staff planning process. The MP on the battlefield are no exception. The MP employ the IPB process as their first step in developing and implementing the PIO function within their commands.

4-65. During the IPB process, the S2/G2 uses all available databases, intelligence sources/products, and related MI disciplines to analyze the threat and the environment. The PIO function supports this process by providing the S2/G2 with collected police, criminal, and combat information that can directly and significantly contribute to the success of the MI effort. In addition to the combat information, the PIO function provides additional information on possible criminal threats and COAs that may support the S2's/G2's IPB process and that can be used by the commander to upgrade the force-protection posture.

POLICE INFORMATION ASSESSMENT PROCESS

4-66. Like the S2/G2 uses the IPB process to continuously analyze the threat and the environment in a specific geographical area, MP leaders use the PIAP as a tool to continuously collect, organize, interpret, and gain access to police/criminal information in support of the IPB process. Criminal trends may have an impact on the tactical scenario, and the PIAP is a method used to consider this threat and its impact on friendly forces. MP leaders cannot use the PIAP as a substitute for the IPB process—the PIAP complements the IPB (see Appendix E).

PASSIVE MODE PIO

4-67. Every MP conducts the PIO function in a passive mode during their normal day-to-day operations and across the full spectrum of military operations. In the passive mode, PIO are not a stand-alone function; as such, they cannot be separated from other MP functions.

4-68. During the performance of MMS, AS, I/R, and L&O functions, the MP develop and exchange information with other organizations in the AO. The MP obtain information through contact with civilians, NGOs, IHOs, local and HN police, multinational police, and other security forces. If the MP receive, observe, or encounter information (police, criminal, or combat) while performing these functions, they will immediately submit a report to relay information up the chain of command. This report may be in the form of a size, activity, location, unit, time, and equipment (SALUTE) report; a spot report (SPOTREP); or another appropriate report. When the higher echelon (brigade, battalion, or PM) receives this information, it is simultaneously integrated into the ongoing IPB/PIAP and forwarded to the higher echelon S2/G2 (see Figure 4-1, page 4-16).

4-69. If police/criminal information is obtained, the MP—

- Update the previous police/criminal estimates provided to the S2/G2.
- Identify new or potential criminal threats or trends in the AO.
- Consider recommending that the supported commander upgrade the force-protection level.
- Notify adjacent units of the potential criminal threat that may affect their forces.
- Consider reprioritizing MP support to the identified threat area.
- Share information with HN/local police and other agencies.

4-70. If combat information is obtained, the MP—

- Forward the information to the higher headquarters S2/G2.
- Forward the information to the MP chain of command, integrate it into the MP's IPB process and, if necessary, take appropriate action.
- Notify the adjacent unit of the potential threat that may affect their forces.

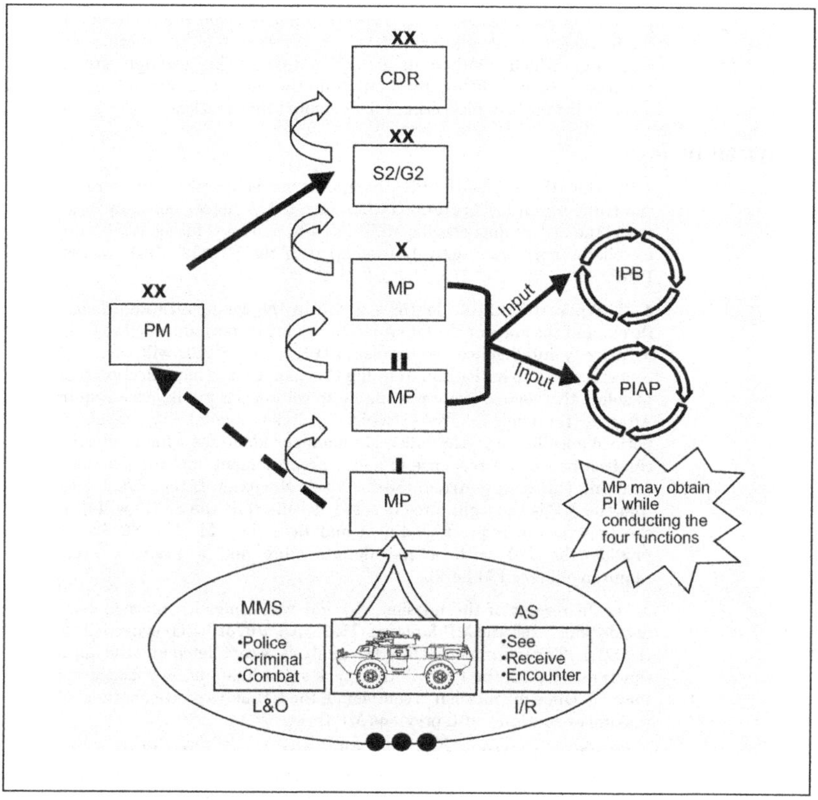

Figure 4-1. PIO—Passive Mode

> During Operation Uphold Democracy, an MP team was conducting a TCP as part of a cordon and search operation in Port-au-Prince, Haiti. While performing the task, two civilians approached the MP team informing them of criminal activity in the neighborhood. During the interview, the MP team prepared a sketch of a house and surrounding areas. The team also obtained information describing the criminals and their weapons. Recognizing that the criminal activity was in fact the actions of a political/mercenary group named FRAP, the MP team radioed the platoon leader and forwarded the field interview to higher headquarters. Two days later, a unit from the 10th Mountain Division raided the house, capturing weapons, ammunition, and equipment.

4-71. The preceding vignette demonstrates the MP performing the PIO function in the passive mode. The MP team received the information while conducting a TCP and submitted it through the appropriate chain of command, which resulted in an action taken. This example stresses the importance of submitting information up the chain of command regardless of whether it may be police, criminal, or combat information.

ACTIVE MODE PIO

4-72. The MP perform the PIO function in the active mode and across the full spectrum of military operations when directed by higher headquarters. In this mode, the MP conduct specific MMS, AS, I/R, and L&O missions with the intent to collect information actively in support of the S2's/G2's IPB process or the PIAP.

4-73. When the S2/G2 identifies a gap in the command's knowledge of the threat and the current threat situation, it may be recommended to be included as priority intelligence requirements (PIR). The S2/G2 will then develop a collection plan to assist him in filling this gap. Part of his collection strategy is to select the best collectors available to cover each intelligence requirement. After a thorough analysis (which includes availability, capability, and performance history), the collection manager identifies which collection asset can best be used in response to a given requirement, and the Operations and Training Officer (US Army) (S3)/G3 tasks the asset. If the S2/G2 determines that the MP is the right force to serve as collectors, the S3/G3 will go through the appropriate request channels and task the PM. The S2/G2 will then provide the PM with a specific guideline and a prioritized collection requirement (see FM 34-2).

4-74. On receipt of the mission, the PM will conduct a mission analysis to decide which specific MP function (MMS, AS, I/R, or L&O) is needed to satisfy the S2's/G2's requirement. Once the analysis is completed and the appropriate function selected, the PM will then task subordinate units with the collection mission. Once the mission is completed, the PM may receive another collection tasking or continue with previous MP tasks.

4-75. When the MP commander or the PM conducts the PIAP, he may also encounter a police/criminal information gap. This gap may become the MP commander's police/criminal information requirements (PCIR). If the gap cannot be filled with available data from the S2/G2, the CID, the HN, and other agencies, the MP commander/PM may task subordinate MP units or request CID assistance to support the collection effort. On receipt of the mission, the MP will then conduct a mission analysis and decide which specific MP function (MMS, AS, I/R, or L&O) is needed to satisfy the PM requirement. Once the mission is completed, the MP may receive another collection tasking or continue with previous MP tasks (see Figure 4-2, page 4-18).

4-76. Since any soldier on the battlefield can report police or criminal information, the MP commander or the PM must constantly coordinate and communicate with the S2/G2, PSYOP, and other agencies to obtain information that could be of MP/CID interest. This constant coordination is a key factor for ensuring that the MP/CID has visibility over the police/criminal information that is reported through non-MP channels.

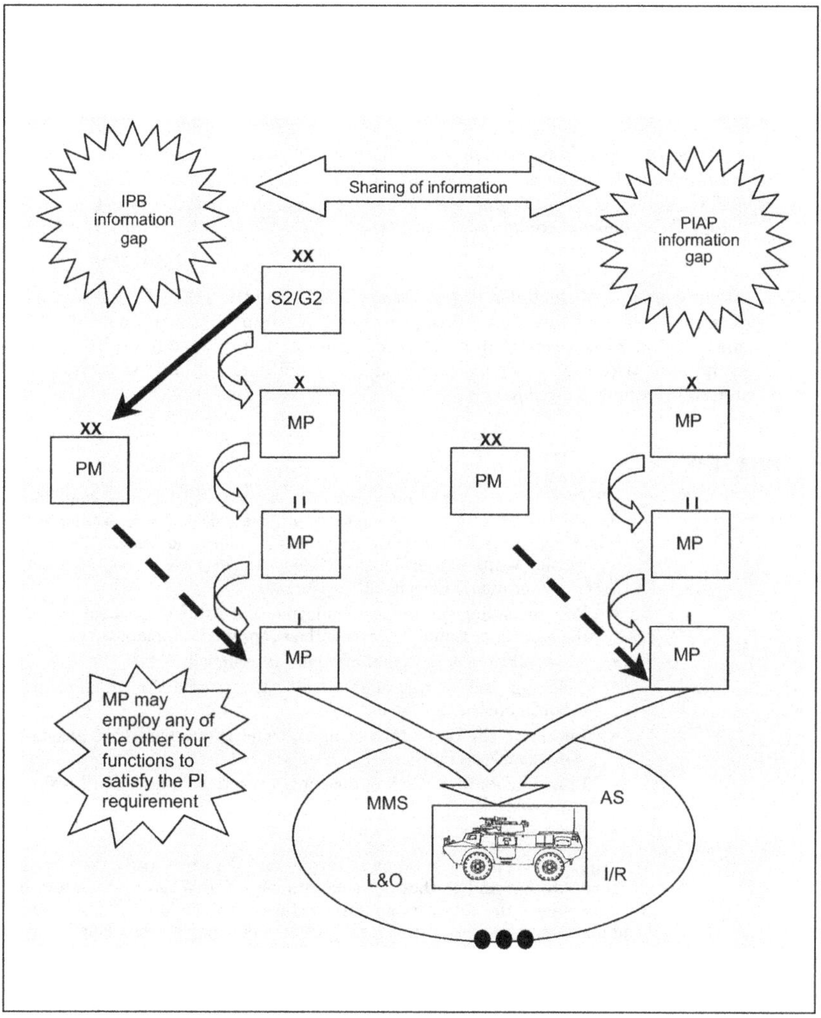

Figure 4-2. PIO—Active Mode

Chapter 5

Military Police Support to Echelons Above Corps

MP units supporting EAC perform combat, CS, and CSS operations. Like the MP supporting corps and divisions, MP units supporting EAC units perform the five MP functions based on available assets and the supported commander's needs.

USAMPS

MP support to EAC includes support to the ASCC and the TSC. The ASCC is responsible for Army Title 10 requirements in support of a combatant commander. This support includes recruitment, organization, supply, equipment, training, servicing, mobilizing, demobilizing, administration, and maintenance functions.

OVERVIEW

5-1. The ASCC may also be responsible for significant DOD- and combatant-commander-designated Army support to other services. As the senior Army commander in the AOR, the ASCC commander tailors and designates ARFOR to accomplish operational-level tasks while conducting major land operations. The ASCC's operational responsibilities include—

- Recommending the proper employment of Army-component forces to the joint-force commander or to the subunified commander.
- Accomplishing operational missions as assigned.
- Selecting and nominating specific Army units for assignment to subordinate theater forces.
- Informing the combatant commander of the Army's CSS effects on operational capabilities.
- Providing data to the supporting operations plans (OPLANs) as requested.
- Ensuring signal interoperability.

5-2. The ASCC provides administrative and logistics (A/L) services to assigned ARFOR and to those of subordinate JFCs. When appropriate, the ASCC delegates the authority for support tasks to a single subordinate Army headquarters. In major operations, the TSC (along with other EAC support commands) would be the ASCC's lead organization for planning, coordinating, executing, or providing required support functions (see FM 100-10).

5-3. The TSC is the senior Army support organization in a theater. Its commander reports to the ASCC or ARFOR commander. The TSC normally operates at the operational level of CSS with links to the strategic and tactical levels. Unity of command is the critical element that the TSC brings to the fight. The TSC is a multifunctional organization that centralizes the command, control, and supervision of support functions at EAC as directed by the ASCC or ARFOR commander. The TSC's mission is to maximize throughput and follow-on sustainment, including all CSS functions, of ARFOR and other designated supported elements. The TSC is capable of synchronizing logistics and other support operations for the ASCC. It provides area support to EAC units in the COMMZ and sustainment support to tactical forces. This support may include supply, procurement, property disposal, maintenance, transportation, field services, health services, civil-military affairs, MP support, engineer support, religious support, finance support, and personnel and administrative services.

5-4. Units and commands requiring support coordinate with the TSC support-operations staff to secure their initial support, to reestablish support, or to resolve support problems. In a fully developed theater, the TSC coordinates with a corps support command (COSCOM) for support of combat forces, although direct coordination with a division support command (DISCOM) is sometimes necessary. The TSC, augmented by a rear operations center, is also responsible for security operations as directed by the ASCC/ARFOR commander (see FM 63-4).

MILITARY POLICE SUPPORT

5-5. MP support to EAC units is provided through an array of multifunctional MP units. The nature of the operation, METT-TC, and the requirements of the supported commander will determine which type of MP unit is appropriate to augment, assign, attach, or place under OPCON to an EAC unit. The types of MP units that support EAC include CS, I/R, CID, and L&O teams (such as MWD or customs teams).

THE MP BRIGADE (CS)

5-6. The MP brigade (CS) provides MMS and AS to extended LOC within the COMMZ. These supply corridors include ports, inland waterways, railways, pipelines, airfields, MSRs, and MSR critical points. The MP support the users of the COMMZ's LOC by aggressively patrolling the area along the LOC. They play an important role in securing rear areas by performing combat operations against the threat. When properly augmented, the MP brigade headquarters may serve as the TSC's/ASCC's TCF headquarters. The MP provide MMS on the COMMZ MSRs leading into the corps's rear area. The MP implement the plans of HN and US staff elements to control the forward movement of combat resources along the LOC.

5-7. If resources are available, the MP brigade (CS) provides escorts to move US noncombatants (if present) from AA points to theater embarkation terminals. Until the MP brigade (I/R) arrives in theater, the CS MP units also perform EPW, confinement, and other operations normally performed by the MP I/R units.

5-8. The organization of an MP brigade (CS) supporting EAC includes the following:

- A brigade headquarters and headquarters company (HHC).
- Up to six MP battalions (each with up to six companies).

- Numerous L&O detachments and MWD teams.

5-9. Additionally, the ASCC's PM or commander may attach or direct OPCON of customs teams to the MP brigade (CS). Battalion and company organization in the MP brigade is the same as that in the corps MP brigade; however, METT-TC determines the number of battalions and companies. The MP brigade (CS) has additional MP companies to provide security for EAC-specific units/activities such as—

- A unified command.
- An ASCC and TSC headquarters.
- LOC seaports, airfields, and railways.
- EAC ammunition storage areas.
- EAC petroleum terminals and pipelines.

5-10. While the corps MP brigade (CS) MWD teams are employed in a GS role, the MP brigade (CS) MWD teams are employed to augment seaport security and to conduct inspections of postal items to detect explosive materiel and narcotics.

5-11. The C^2 within the MP brigade (CS) is consistent with that in any Army brigade. The MP brigade commander works directly for the EAC commander, the battalion commander works for the brigade commander, and the company commander works for the battalion commander. However, this usual C^2 relationship may be altered briefly (based on METT-TC) to enhance the overall EAC combat capability for responding to a Level II threat. For example, MP units operating within an ASG's AO may be under the OPCON of the ASG's rear-area operations center (RAOC), which directly tasks MP units responding to Level II threats. The same is true for placing MP units under the OPCON of the EAC's TCF headquarters for responding to Level III threats. Any conflict in mission priorities is resolved through MP C^2 channels.

5-12. The MP brigade commander is both the MP brigade commander and the EAC's PM. He employs his assets according to METT-TC and the commander's concept of operations. Factors affecting his employment of MP assets include the—

- Nature of the operation (joint, combined, or multinational).
- HN's ability to provide MP-related support (such as port security).
- Custody and location of EPWs/CIs during internment operations until I/R units arrive in theater.
- Number of kilometers of the MSR in relationship to movement-control requirements.
- Number and kinds of critical facilities.
- Number of HRP requiring close-in security.
- HN's ability to control the civilian populace, refugees, and dislocated civilians.
- Supply distribution strategy.
- Risk acceptance and threat in the AO.
- Communications requirements (such as using teams as relays).

5-13. Whenever possible, the MP brigade's AO coincides with the territorial responsibility of the supported command. The MP brigade commander assigns the MP battalion's AO by the above factors as well as by METT-TC. For example, the AOR for an MP battalion may be a large population center of larger geographical areas in which CSS complexes and MSRs are located. But as employment factors and the commander's needs change, so will the MP's AOR. The MP brigade commander must move and tailor his forces to meet the current and projected mission requirements. Unlike many other EAC assets, MP units require 100 percent mobility to shift AOs frequently and rapidly. The following vignette depicts the required MP flexibility to support EAC operations:

> During Operation Desert Shield/Storm, MP units were supporting and moving units throughout their AOs at a such an extraordinary rate that many of them had to relocate their headquarters multiple times just to keep pace with their changing AO.

5-14. Most EAC MP assets are employed along LOC and around areas of high troop concentration. Few EAC MP are dedicated to support fixed commitments (such as ports, air bases, and headquarters [discussed in paragraph 5-8]). When supporting fixed commitments, MP units provide a mobile security screen, and they man static positions when directed or when based on METT-TC. The MP brigade (CS) may have to plan for and actually perform the evacuation and internment of EPWs/CIs and the confinement of US military prisoners until the MP brigade (I/R) arrives.

THE MP BRIGADE (I/R)

5-15. US policy requires that all persons held in Army custody be accorded humane care and treatment from the moment of custody to their final release. The policy applies to detained or interned civilians as well as to EPWs and confined US military personnel. This policy is equally binding on all US troops (see FM 19-40).

5-16. The ASCC or the TSC supports US laws, regulations, policies, and international agreements by providing personnel, administrative, morale, internment, resettlement, and confinement services for the TO. The TSC's MP brigade (I/R) in the TO provides this support. However, since most I/R units are in the reserve components, the initial I/R operations (as mentioned above) may have to be conducted by the MP brigade (CS). Once the I/R unit arrives in the AO, it is responsible for—

- Providing firm but humane control of EPWs/CIs and dislocated civilians.
- Coordinating with HN personnel, military territorial organizations, civilian police authorities, NGOs, private volunteer organizations, and US federal agencies on matters pertaining to I/R operations.
- Performing C^2 operations for all I/R units.
- Controlling, employing, and releasing EPWs/CIs as set forth by the Geneva convention and other international laws and by the UN and other governmental bodies.
- Handling US military prisoners.

5-17. In a mature theater in which large numbers of EPWs are captured, the EPWs' requirements may exceed the capacity of the MP brigade (I/R). In this instance, an I/R command is established. An MP command (I/R) has two or more MP brigades (I/R) and will normally be assigned to the ASCC. When the MP command (I/R) assumes OPCON of the MP brigades (I/R) from the lower echelon, it assumes that echelon's I/R mission.

5-18. If the US decides to transfer captured EPWs/CIs to the HN or to another nation, the US must ensure that the nation is a party to the Geneva convention and is willing and able to comply with the convention. In this case, the number and type of I/R MP units required for processing and retaining EPWs/CIs before the transfer is based on agreements and on METT-TC. Additionally, the MP brigade (I/R) is assigned I/R teams that are located at the processing and transfer points and at the HN or third-country EPW camps. The MP brigade (I/R) liaison team will supervise these dispersed teams to ensure that the HN or the third country provides adequate care and security of US-captured EPWs/CIs and that accountability is maintained according to the Geneva convention.

5-19. The MP brigade (I/R) HHC is the C^2 element for the brigade's assets. It consists of the following elements:

- A brigade command section.
- A company headquarters.
- An Adjutant (US Army) (S1).
- An S2.
- An S3.
- A Supply Officer (US Army) (S4).
- A Civil-Affairs Officer (US Army) (S5).
- A Communications Officer (US Army) (S6).
- Finance and accounting.
- Medical operations.
- Public affairs.
- A unit ministry team.
- An SJA.
- An inspector-general (IG) section.

5-20. Other brigade assets may include the following:

- An MP I/R BLD.
- An I/R information center.
- An MP escort-guard company.
- An MP I/R battalion headquarters.
- MP I/R (EPW/CI) detachments.
- MP I/R (confinement) detachments.
- MP guard companies.
- MWD teams.
- Processing squads, processing liaison teams, camp liaison teams, and evacuation teams (all as required to support EPW transfer or to conduct an out-of-theater evacuation).

5-21. The MP brigade (I/R) subordinate units are employed most often in the COMMZ near CSS facilities and are placed near sea, air, and rail terminals. They receive, process, and intern EPWs/CIs; confine US military prisoners; or assist in the resettlement of refugees or dislocated civilians.

5-22. The MP brigade (I/R) escort-guard company supports the evacuation of EPWs/CIs from the CZ. The company has a company headquarters and four platoons. The escort-guard company requires sufficient mobility to transport MP personnel to the CZ and to operate mobile teams while escorting the EPWs/CIs. The MP go forward to the corps's holding area to take custody of the EPWs/CIs. They may go forward to division collecting points, if distances and resources permit. Using any means of available transportation, the MP ensure that the EPWs/CIs are quickly evacuated to MP battalions (I/R) in the COMMZ. Close coordination with the EAC and corps movement-control centers and the corps MP brigade is required to ensure that transportation assets returning to the COMMZ are employed to evacuate EPWs/CIs from the corps's holding area. Walking wounded EPWs/CIs are evacuated by the same means as other EPWs/CIs, while litter patients are evacuated through medical channels. Guarding EPWs/CIs while in the MEDEVAC channels and during their hospitalization is not an MP mission; therefore, there is not an MP force structure to support this mission. In most instances, the impact of having the MP perform this mission causes trade-offs in missions for which they are responsible.

5-23. The theater MP brigade (I/R) and out-of-theater MP brigade (I/R) subordinate units will evacuate EPWs to internment sites within CONUS (if directed). The theater brigade structure is based on the projected capture rate over time and available out-of-theater transportation assets (frequency and capacity). The out-of-theater brigade structure is based on the total EPW/CI population, the number of internment sites, transportation nodes, and escort requirements. Theater escort-guard MP move the EPWs/CIs to the seaport and aerial port of embarkation (SPOE/APOE). The escort-guard MP assigned to the out-of-theater brigade escort the EPWs/CIs from the theater ports of embarkation (POEs) to the out-of-theater internment sites. The out-of-theater brigade is assigned an I/R evacuation detachment, which is employed at and coordinates the evacuation from the theater POEs, through the out-of-theater ports of debarkation (PODs), to the out-of-theater facilities.

THE MP BATTALION (I/R)

5-24. The MP battalion (I/R) is a modular organization and can be configured to operate internment facilities for EPWs/CIs, confine US military prisoners, or resettle dislocated civilians. When performing EPW/CI internment operations, the MP brigade (I/R) has up to 7 MP battalions (I/R); when augmented with the appropriate number of BLDs, it has up to 21 MP battalions (I/R). The ASCC, the TSC, and the MP brigade (I/R) must consider that the requirement to establish an MP battalion (I/R) internment facility is resource intensive. Therefore, MP I/R units, other supporting units, supplies, and equipment for the EPWs/CIs should arrive in theater ahead of the projected EPW/CI arrival at the internment facilities. Early arrival should be based on the time required to establish fully operational facilities (construct and man) and resupply operations before the EPWs/CIs arrive.

5-25. The MP battalion (I/R) has a command section, a company headquarters, and various staff sections. The staff sections provide the core battalion-level capabilities to conduct internment operations. The modulated design expands as the EPW/CI population increases. The battalion is assigned up to four detachments, two guard companies and, if needed, a processing squad. When fully operational, an MP battalion (I/R) operates an enclosure capable of interning 4,000 EPWs/CIs. The battalion mission centers on eight 500-man compounds. The battalion operates the compounds in close proximity to maximize its resources for the security and internment of the EPWs/CIs. This includes the resources needed to employ EPWs/CIs as a labor force according to the provisions of the Geneva convention.

5-26. Each MP detachment (I/R) (EPW/CI) operates two 500-man compounds and provides augmentation to the battalion staff sections to support 1,000 EPWs/CIs. Each guard company is capable of providing security for 2,000 EPWs/CIs. The guard company has a company headquarters and three platoons. The guard company requires sufficient mobility and communications to support routine battalion missions. While minimum mobility and communications is required to support EPW/CI internment operations, on-site guard personnel must often move considerable distances guarding labor groups performing work projects throughout the COMMZ.

5-27. The MP (EPW/CI) processing squad is capable of processing eight EPWs/CIs per hour and includes interpreters to support the processing. If processing squads are required to augment MP battalions (I/R), the operational requirements will be based on METT-TC.

MP-Battalion Resettlement Operations

5-28. The basic organization used for EPW/CI internment is used for resettlement operations. The primary mission-focus change is from guarding EPWs/CIs to protecting and controlling dislocated civilians. As such, an MP battalion (I/R) with four detachments and two guard companies is capable of supporting 8,000 dislocated civilians. However, the MP battalion (I/R) may require augmentation to conduct L&O operations associated with the resettlement. Augmentation may include the full scope of PM functions (operations, investigations, physical security, MP-station operations, and patrols) and civil affairs.

MP-Battalion Confinement Operations

5-29. When configured with the MP detachment (I/R) (confinement), the MP battalion (I/R) is capable of confining US military prisoners. The MP detachment (I/R) (confinement) provides trained corrections and support personnel required for confinement operations. As with the EPW/CI configuration, the modular confinement structure expands as the US prisoner population increases. With three confinement detachments and three guard companies, the MP battalion (I/R) has a maximum capacity of handling 1,500 US prisoners. Generally, only one MP brigade battalion (I/R) is configured for confinement, but the actual number will depend on the number of US prisoners requiring confinement within a theater. While theater policy for confinement operations remains with the ASCC commander, it is the MP brigade (I/R) that executes the mission.

5-30. When possible, soldiers awaiting trial remain in their units unless reasonable grounds exist to believe that they will not appear at the trial, the pretrial hearing, or the investigation or that they will engage in serious criminal misconduct. Under either of these two pretrial confinement instances, the commander must also reasonably believe that a less severe form of restraint (such as conditions of liberty, restriction in lieu of apprehension, or apprehension) is inadequate. When these circumstances exist and other legal requirements are met, US military personnel may be placed in pretrial confinement under the MP's direct control. Commanders may choose to establish field confinement facilities within their AO. However, corps and division MP companies have the expertise to operate only a field detention facility for a limited period of time. These units cannot operate a confinement facility and have neither the resources nor the capability to operate such a facility on an extended basis. Therefore, all confinement is consolidated in the COMMZ whenever possible.

5-31. All assets of the MP battalion (I/R) (confinement) are employed to detain, confine, sustain, and protect US prisoners. As with the battalions conducting EPW/CI and resettlement operations, the battalion conducting confinement operations is generally located in the rear of the COMMZ, near logistics and transportation support. This allows US prisoners to be moved as quickly as possible from the corps's area to the COMMZ's confinement facility. Movements of US prisoners from the COMMZ to CONUS will be according to DA policy.

Chapter 6

Military Police Support to Corps

Corps MP support their command by performing the MP functions critical to the success of their tactical commander's concept of operations.

USAMPS

Corps are the largest tactical units in the US Army. They are the instruments by which higher echelons conduct operations at the operational level. Higher headquarters tailor corps for the theater and the mission for which they are deployed. They contain organic combat, CS, and CSS capabilities to sustain operations for a considerable period of time. Corps are capable of operating in a joint and multinational environment, providing C^2 for up to five divisions and covering up to 35,000 square kilometers.

OVERVIEW

6-1. MP support to a corps is provided by an MP brigade (CS) assigned to each corps. The MP brigade provides combat, CS, and CSS throughout the corps's AO. However, subordinate MP units are not assigned to subordinate corps units. Instead, the MP brigade commander gives them an AO based on the corps commander's concept of operations. When possible, MP battalion AOs coincide with those of the CSG RAOCs.

6-2. The corps MP provide combat power within the command's rear area. They perform combat operations to counter Level II forces and to support the defeat of Level III forces. When properly augmented, the MP brigade may serve as the corps's TCF. The corps MP also provide a critical link between MP operations in the division and in the COMMZ. The corps MP support division commanders by helping the division MP conduct sustainment operations. The corps MP coordinate with the division MP for mutual support.

THE MILITARY POLICE BRIGADE (COMBAT SUPPORT)

6-3. The MP brigade (CS) supporting a corps contains a brigade headquarters, up to six MP battalions (CS), numerous L&O detachments, and MWD teams. Each MP battalion (CS) has up to six MP companies (CS). As with the EAC's MP brigade (CS), the number of battalions and companies is determined by METT-TC. The corps MP brigade (CS) has additional companies to support each division and to provide security for the corps and COSCOM headquarters and corps ammunition storage areas (see Figure 6-1, page 6-2).

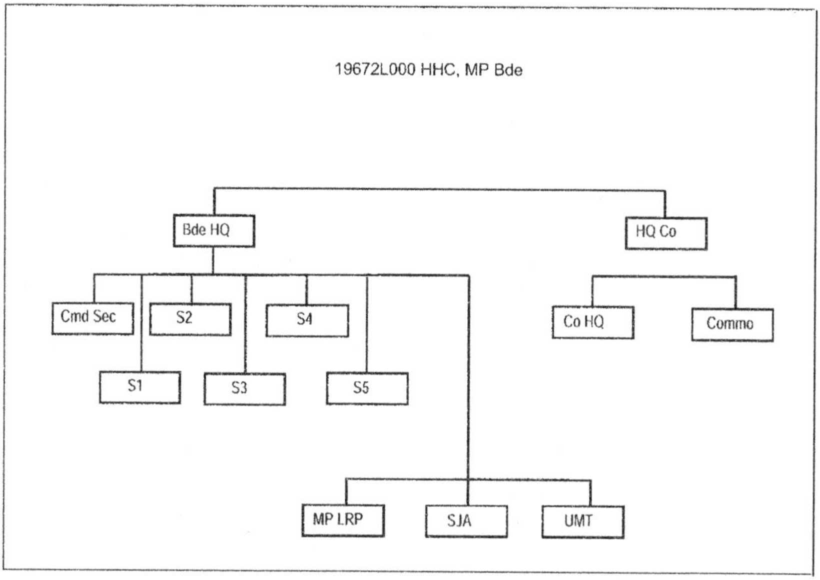

Figure 6-1. Military Police and USACIDC Organizations

6-4. The MP brigade HHC provides C^2 and A/L support to the brigade. The brigade HHC consists of a company headquarters and a brigade headquarters that contains the commander's immediate staff. The staff officers supervise the brigade's major organizational elements, including the—

- S1, S2, S3, S4, and S5.
- MP long-range plans (LRP) section.
- Communications section.
- SJA.
- IG.
- Public affairs.
- Unit ministry team (UMT).

6-5. The MP brigade (CS) command sergeant major (CSM) requires mobility and communications capabilities to execute his duties and responsibilities. The MP LRP section works with the corps G3 plans element, operating out of the corps main CP.

6-6. The MP battalion headquarters and headquarters detachment (HHD) provides C^2 for MP CS companies and any other assigned or attached MP elements. The battalion HHD consists of a detachment headquarters and a battalion headquarters that contains the battalion commander's staff. The staff officers supervise the major organizational elements, including the—

- S1, S2, S3, and S4.
- Communications and support sections.
- UMT.

6-7. As with the MP brigade (CS), the battalion CSM requires mobility and communications. The support section is vital to an MP battalion commander's ability to sustain his widely dispersed assigned or attached units during the performance of all five MP functions, primarily the MMS and AS functions.

6-8. The MP company (CS) provides support to an assigned AO. The company has a company headquarters, an MP operations center, a combat-medic section, and four platoons. The company headquarters provides maintenance, supply, communications, mess, and medical support to the unit. An MP operations center supports the unit's operation, conducting and planning for all five MP functions. The MP operations center includes three MP teams as the company-level response force.

6-9. The MP L&O detachments provide support to an assigned AO. The headquarters provides A/L support. The operation team plans and supervises desk operations, the traffic-accident and MP-investigation teams, and the force-protection teams. As with the EAC MP brigade (CS) detachments, requirements are based on the population supported and on METT-TC.

6-10. The number of assigned MWD teams is significantly less than those assigned to the EAC MP (CS) or (I/R) brigades. A kennel master, five explosive/patrol teams, and four narcotics/patrol teams are normally assigned to the corps MP brigade (CS). They are employed based on METT-TC.

COMMAND AND CONTROL

6-11. The C^2 in a corps MP brigade (CS) is consistent with that of any Army brigade. The MP brigade commander commands the brigade and all attached personnel. Battalion commanders work for the brigade commander. The MP company commanders receive their orders and work for their respective battalion commanders and direct their platoon leaders according to mission requirements.

6-12. When corps MP assets are sent to augment the division PM or the MP company, they are placed under the OPCON of the division PM for as long as the unit is needed in the division area. This command relationship is applicable to company-size organizations and smaller. If an MP battalion (or larger MP unit) augments the division, then the division PM will not exercise OPCON of that MP unit. The MP battalion will be placed OPCON as designated by the division commander but not under the division PM. One L&O detachment is normally attached to the supported division and placed OPCON to the division PM. However, MP leaders at each level establish an MP C^2 relationship based on METT-TC and the supported commander's needs.

EMPLOYMENT

6-13. Each corps MP brigade commander employs his assets according to METT-TC, the needs of the forces operating in his AO, and the priorities of the corps commander. Few MP assets in the corps area are employed to support fixed commitments. Instead, based on a broad consideration of the enemy and friendly situations, the corps MP are employed to support friendly forces engaged in combat, CS, and CSS operations. Since MP forces are dispersed throughout the corps area, the concentration of US forces, the location and vulnerability of critical sites, and the number of kilometers of the MSR to be controlled influence the designation of MP assets in the AO.

6-14. The MP brigade commander establishes his MP concept of operations based on the corps commander's concept of operations. His successful employment of MP assets depends on his foreseeing where the battle will be rather than where the battle is. Based on the rear-area IPB and PIAP, the MP brigade commander allocates and shifts resources to ensure the accomplishment of priority missions. This ensures the continuous support and forward sustainment of combat units and the safety of CSS units operating in the corps rear and sustainment areas.

6-15. To support the MP brigade commander's planning, the MP at brigade headquarters operate from several locations. Most of the staff locates in the vicinity of the corps rear CP where they can interface with the corps staff responsible for planning and executing rear operations. The control element of the brigade headquarters must be located where it can command and control its subordinate units. The brigade commander and his staff decide the best place to locate this element. The planning element of the headquarters locates near the corps main CP where it can interface with the corps commander's coordinating, special, and personal staffs. From there they monitor MP operations, integrate MP support with the corps plans cell for future operations, and learn the enemy situation through the G2's threat analysis almost immediately. The detailed information on rear-area activities and operations provided by the corps staff enhances the accuracy of the MP LRP.

6-16. The brigade S3 section provides the day-to-day planning and execution of an MP mission. The section provides a responsive CP that can relocate frequently throughout the AO. The S3 will normally provide liaison personnel to the corps rear CP, selected COSCOM units, interagencies, or other headquarters (based on mission requirements). When possible, battalion AOs coincide with the AOs of the CSG RAOCs. The MP brigade commander usually tailors battalions' AO boundaries to ensure responsive and flexible support across the corps's AO. He pays particular attention to the LOC behind the most heavily committed division and the critical bases and facilities in that area. He also ensures that the MP are available to respond quickly to combat operations throughout the entire corps rear area or during sustainment operations.

6-17. The MP brigade commander, coordinating with his battalion commanders, locates the MP companies where they can provide combat and CS power throughout the corps's AO. He bases his decision on the—

- Number and composition of urban areas.

- Location of CS and CSS units.
- Location of critical facilities (such as the headquarters, ammunition storage areas, and airfields).
- Expected threat.
- Level and frequency of support needed by the commander.
- Current and projected tactical situation.
- MSR network, including choke points and critical bridges and tunnels.
- Number of supported divisions and requirements.

6-18. The prioritization of MP missions is especially crucial during the early stages of the deployment when it is unlikely that an MP brigade commander will have a full complement of CS companies. Augmenting the division MP company with corps MP assets may not be possible at that time. Until it is, the MP brigade commander must locate corps MP assets to help meet the needs of the division while fulfilling the needs of the corps.

6-19. Like the brigade commander, battalion commanders plan the employment of their companies and platoons using METT-TC. Certain corps needs are constant. One MP company is allocated to provide security for the COSCOM, and one MP company is allocated to provide security for the corps's main CP. One platoon from that company may be used to secure the corps tactical CP or the jump CP. One or more platoons will also help secure the corps's rear CP. The number of MP assets allocated for a corps-level EPW/CI holding area and escort is based on METT-TC. However, a minimum of one platoon is dedicated to operate the corps's EPW/CI holding area and a minimum of one platoon per division is allocated for evacuating EPWs/CIs from division collection points. Additional MP assets may be allocated to provide security for the corps's ammunition storage area and ASPs supporting the divisions. The MP battalion commander places his companies where—

- MP assets support the brigade commander's concept of operations.
- The MP can support troop concentration, bases and base clusters, road networks, and critical areas.
- The MP can aggressively patrol critical terrain and monitor LZs and DZs to detect or deny enemy interference.
- The MP can respond to Level II threats.
- The MP can support the movement of combat resources throughout the AO.
- The MP can remove EPWs/CIs from division collection points.
- The MP can influence stragglers, refugees, and dislocated civilians.

6-20. Battalion commanders may choose to place a company behind the division rear boundary. This can help to ease the coordination between the corps MP and the division MP.

Chapter 7

Military Police Support to Division

Division MP provide essential support to the forward tactical commander. The MP provide a flexible, mobile, and lethal force ready to be employed in combat or CS operations.

USAMPS

The division is a large Army organization that trains and fights as a tactical team. Largely self-sustaining, it is capable of independent operations. The division is a unit of maneuver organized with varying numbers and types of combat, CS, and CSS units. Divisions fall mainly into two categories—heavy or light. The first category includes heavy divisions, their variants, and the Division XXI digitized heavy division. The second category includes the light infantry, airborne, and air-assault divisions. Each division can conduct operations over a wide range of environments. The success of Army operations depends on the success of its divisions.

OVERVIEW

7-1. In corps operations, divisions are normally comprised of three maneuver brigades, each with up to nine maneuver battalions, artillery battalions, aviation battalions, and supporting CS and CSS units. Divisions perform a wide range of tactical missions and are self-sustaining for limited periods of time. The corps augments divisions as the mission requires. All divisions must be able to deploy and conduct offensive, defensive, stability, and support operations. Airborne and air-assault divisions must be able to conduct forced-entry operations. Divisions may be part of a JTF or serve as the ARFOR headquarters.

7-2. MP support to a division is provided by an organic MP company that includes the PM and his staff (assigned to each division) and is augmented by the corps MP brigade. The division MP company provides support in the DRA and in the maneuver brigade rear area. It is fully mobile in order to relocate frequently under short notice. While division MP companies are capable of performing all five MP functions, they must be augmented by corps MP/CID to fully perform all five MP functions throughout the division's battle space. The division PM must receive at least one corps MP company for GS missions in the DRA, one L&O detachment for L&O missions, and one CID detachment for criminal investigations and LOGSEC. In a mature theater (or based on METT-TC), the division PM can expect two corps MP companies. In addition, the division PM may request a corps MP battalion headquarters to provide C^2 to corps MP units operating in the DRA or to perform missions as a TCF. Because the need for MP support exceeds division organic assets (and many times exceeds augmenting corps MP assets), careful planning of MP employment is essential.

DIVISION MILITARY POLICE COMPANY

7-3. The organization of the division MP company is set by specific tables of organization and equipment (TOE) and is designed to support a specific type of division. Regardless of the type of division, the assets in a division MP company enable a PM and his supporting staff to supervise the performance of all five MP functions. The PM coordinates the employment of the MP assets in the division's AO through his PM section. The division MP company contains a headquarters, a PM section, a medic section, and three to six platoons.

COMMAND AND CONTROL

7-4. The C^2 of the division MP company extends downward from the tactical commander to the division PM. The division PM is the division commander's special staff officer. Under normal circumstances, he works for the chief of staff. The division PM has OPCON of the division MP company and MP assets that have been provided by the corps. The division PM plans and employs all of the organic, assigned, or attached MP according to METT-TC. He anticipates the support needs of the division commander and plans the employment to meet his needs. After learning the division commander's concept of operations and estimating the situation, the division PM decides which MP functions are critical to accomplish the commander's mission. He then recommends the allocation of resources, tasks his assets, and establishes the priority of support.

7-5. When the division MP company is augmented by corps MP platoons, the platoons are placed under the direction of the division MP company commander, who assigns an AO to each platoon. When corps MP assets augmenting the division make up a company, the company is attached to the division. The company then comes under the OPCON of the PM, who assigns it an AO. The augmenting corps MP company is usually given an AO from the division rear boundary forward. However, this assignment is flexible and based on the division commander's needs as set forth by the division PM. The four platoons of that corps MP company perform all five MP functions in that AO. Additionally, the corps MP brigade (CS) L&O detachment supporting a division is attached to the division and OPCON to the division PM.

7-6. The division MP company commander directs the employment of company assets. Through his company headquarters, he provides administrative, maintenance, and logistical support to the PM section and to the company's platoons that are dispersed throughout the division's AO. These sections are mobile to support the platoons without delay despite the extended distances that may be separating them. The company commander is also mobile. He travels the extended distances from one platoon area to another when checking on the status of his soldiers and resolving the problems affecting mission accomplishment. The division HHC provides mess support. Personnel administration is handled by the company headquarters and the DISCOM. The number and kind of assets in a division MP company and the configuration of its platoons and squads are determined by the type of division to which the company is assigned.

NOTE: See the consolidated TOE update that is published yearly for complete details (base and objective) for all MP TOEs.

7-7. At the direction of the division commander, the division band may be available to perform its secondary mission of supporting MP operations. When the intensity of the conflict prevents the band from performing its musical mission, the division commander may direct it to augment the MP company. The band, normally employed as a unit under the direction of the bandmaster, is under the OPCON of the MP company commander while augmenting the MP. When tasked to augment MP assets, band members may be employed to augment MP security at the division main CP or to augment perimeter security at the division EPW/CI collection point. The band is released to perform its primary mission as soon as the tactical situation permits.

MILITARY POLICE EMPLOYMENT

7-8. The employment of division MP companies differs somewhat with the type of division to which they are assigned. In heavy divisions, the division MP companies provide GS to the DRA and DS to the division's maneuver brigades. When the platoon is providing DS to a maneuver brigade, the MP platoon leader coordinates all logistical support with the supported brigade. A platoon's DS relationship is seldom interrupted. However, METT-TC may require weighing MP support to the maneuver brigade tasked as the main effort or employing all platoons in a division MP company to function as a unit (such as during division-size river-crossing operations or deliberate breaching operations). In light divisions, the MP provide only GS to the DRA or during sustainment operations. Support exceptions are addressed under MP support to airborne and air-assault divisions, discussed later in this chapter.

7-9. In the division (where flexible support of an austere force is crucial), the division PM must have a clear understanding of situational awareness. To obtain current information for projecting MP needs in the division area, he must be mobile and be able to conduct split-cell operations. The assets available to the PM include the division MP company and at least one corps MP company. Corps augmentation is required for sustained operations and for special operations such as river crossings, dealing with dislocated civilians, and refugee internment or resettlement. The division PM coordinates with the corps PM and the MP brigade or CID commanders for—

- Evacuating and guarding EPWs/CIs from division to corps.
- Providing law-enforcement assistance to HN forces in the division's AO.
- Providing corps augmentation for the division's AO, convoy security, LOC security, AS, R&S, L&O, and other missions. These other missions may include augmentation for security of the division main CP, ASPs, and other critical facilities.
- Integrating, sharing, and exchanging police intelligence between corps and division MP elements.
- Providing CID support.

SUPPORT TO THE HEAVY DIVISION

7-10. The Army's armored and mechanized infantry divisions (normally referred to as heavy divisions) provide mobile, armor-protected firepower that is normally employed for their mobility, survivability, lethality, and psychological effect (shock) on the enemy. These divisions destroy enemy armored and mechanized forces, and they can seize land areas and secure key terrain. Because of their strategic lift requirements, heavy divisions are slow to deploy from home staging bases into an AO. They have high consumption rates of supplies and have limited use in restrictive terrain. These capabilities and limitations are key factors in planning effective and efficient MP support throughout the division's AO.

7-11. In the heavy divisions, the PM section is organized to support split-cell operations at the main and rear CPs. The PM must be mobile to ensure that he is fully aware of the current status of critical MP operations. Therefore, the deputy PM (DPM) locates in one of the PM cells and—

- Handles the section's routine operations.
- Monitors ongoing division operations and MP support.
- Helps provide long-range planning and interface with the primary division staff.
- Forwards PM taskings to the division MP company commander.

7-12. The DPM and the operations sergeant normally set up operations at the division main CP where they can coordinate requirements with the division staff personnel. The company headquarters is initially located near the division rear CP in the division support area (DSA). Once augmentation arrives from the corps, the headquarters relocates with one of its GS platoons operating behind the brigade rear boundary. However, this location could be changed based on METT-TC, the supported commander's needs, and the company commander's idea of where he can exercise better C^2 for his unit.

7-13. The Army of Excellence (AOE) heavy division MP company has six platoons. Three platoons provide support to each maneuver brigade and are designated as DS. The other three platoons are designated as GS platoons. One MP platoon provides security for the division main CP; one provides security for the division's EPW central collection point; and one performs other MP operations within the division rear.

7-14. The GS MP platoons' AOs are configured based on METT-TC and the availability of MP augmentation from the corps. The DS MP platoons' AOs coincide with the supported maneuver brigade's boundary. Each platoon headquarters locates within its brigade's support area or any other area where it can best provide and receive support. To accomplish its mission, each DS platoon requires a minimum of two squads, each with three teams. One squad operates the EPW/CI collection point. The other squads perform MMS and AS operations. All MP platoons are capable of performing all five MP functions. However, performance of these functions is prioritized based on METT-TC and the division commander's concept of operations. The division PM, the company commander, and METT-TC dictate how these platoons should be tasked-organized to accomplish the mission.

SUPPORT TO THE LIGHT INFANTRY DIVISION

7-15. The AOE light infantry division (LID) is one of the most rapidly and strategically deployable divisions. It fights as part of a larger force in war or conducts missions as part of a joint force in MOOTW. Its C^2 structure readily accepts any augmentation forces, permitting task-organizing for any situation. The augmentation required for the division is largely determined by METT-TC. The division's capabilities allow it to exploit the advantages of restricted terrain and limited visibility. It achieves mass through the combined effects of synchronized small-unit operations and fires rather than through the physical concentration of forces on the battlefield. These characteristics are key factors in planning and employing MP assets in support of the LID.

7-16. The LID MP companies are capable of performing all five MP functions. However, their performance of these functions is prioritized based on METT-TC and the division commander's concept of operations. Contrary to the heavy division MP company, the LID MP companies are much smaller. The constrained size of the LID MP companies makes corps MP augmentation crucial to the sustainment of MP operations. Additionally, the LID MP company is the only MP unit with the capability of antiaircraft support through the use of shoulder-fired air-defense weapons.

7-17. The company has three GS platoons to support the division. No platoons are provided to the maneuver brigade. One platoon is normally located in the vicinity of the division main CP so that its resources can help support CP security. Another platoon locates in the DSA and operates the division EPW/CI collection point. The last platoon has an AO configured according to METT-TC and the commander's priority of MP missions. Each GS MP platoon has a headquarters and three squads, each with two teams. The PM section is located in the vicinity of the division main CP. The exact location is based on the current operational status and on METT-TC.

SUPPORT TO THE AIRBORNE DIVISION

7-18. The AOE airborne division can rapidly deploy anywhere in the world to seize and secure vital objectives. It conducts parachute assaults to capture initial lodgments, executes large-scale tactical raids, secures intermediate staging bases (ISBs) or forward operating bases (FOBs) for ground and air operations, and rescues US nationals besieged overseas. It can also serve as a strategic or theater reserve as well as a reinforcement for forward-presence forces. The airborne division can assault deep into the enemy's rear area to secure terrain or to interdict enemy supply and withdrawal routes. It can seize and repair airfields to provide a FOB and airheads for follow-on air-landed forces. It is capable of all other missions assigned to LIDs. The airborne division does not have sufficient armored protection to defeat heavier armored formations at close range. Therefore, engagements with enemy armored formations require special considerations. Antitank weapons in the division compensate for, but do not completely offset, this deficit.

7-19. Airborne divisions conduct operations in two phases—assault and defense. The division accomplishes the assault phase in three stages. First, they deploy; second, they establish the airhead; and third, they extend the logistics base and build their force.

7-20. Like other division MP companies, those assigned to the airborne division are employed to support their division commander's concept of operations. The airborne MP company has four platoons, each having three squads with two teams. Normally, the company headquarters and one of the platoon headquarters colocates with the PM section. The division PM's exact location depends on METT-TC. During the assault phase, the airborne division MP company provides DS to the assault brigade. The remaining platoons provide MMS and AS in the vicinity of the division main CP.

7-21. The nature of airborne operations makes the capture of EPWs likely. Therefore, during the first stage of the assault phase, the priority of MP support is given to EPW operations. After assembling the DZ or LZ, the MP collect EPWs captured during the assault. Combat elements are relieved of EPWs as far forward as possible. In airborne operations, EPWs are held for later movement to a central collection point. During the first stage of the assault, the MP perform limited straggler and refugee control and undertake AS operations, when possible.

7-22. MP support is reevaluated after the airhead is established in the second stage of the assault. The PM takes in consideration personnel and equipment flow, roadways, and security requirements to shift MP support priorities. As the entire operation matures, MP support may expand to include all five MP functions.

SUPPORT TO THE AIR-ASSAULT DIVISION

7-23. The air-assault division executes tactical missions at operational depths to achieve strategic results. It is capable of launching brigade-size air assaults of nearly 4,000 soldiers from either an ISB or a tactical assembly area (TAA). Within 6 hours, this air-assault task force (AATF) can attack an opposing force, occupy and defend key terrain, or establish a FOB (out to 150 kilometers) from which even deeper operations can be executed. Air-assault operations are terrain independent, but they rely on suitable weather and a detailed attrition/assessment of enemy capability—particularly air-defense assets along air corridors and in the objective area. The air assault is preceded by detailed, lethal and nonlethal condition settings, culminating in a comprehensive condition check before execution.

7-24. As the corps AATF unit, the brigade combat team (BCT) task-organizes and habitually trains with both aviation lift and attack (Apache) battalions. An air-assault division MP company has four platoons, each having three squads with two teams. To facilitate operations, air-assault division MP platoons are habitually aligned with each of the three AOE BCTs; however, as with aviation assets, the MP are task-organized to support the main effort's BCT/AATF. When conducting these operations, two MP platoons may be tasked to provide support; one is available to posture and participate in the air assault, and the other conducts AS and MMS operations at the pickup zone (PZ). This second MP platoon may further provide MMS and accompany a ground-assault convoy (GAC) to the objective, or it may remain at the PZ to receive EPWs returning on aircraft from the objective area. The mission of conducting MMS along this vulnerable ground LOC is particularly critical if the AATF objective is a FOB from which further division operations will expand.

7-25. MP participation in the actual air assault competes for heavy lift with FA and air-defense systems, as well as with vital Class V resupply. However, the lethality and versatility of the MP bode well for their employment, and two MP platoons are available to support the brigade air assault as the division's center of gravity.

7-26. When possible, habitually aligned platoons remain with their brigades, and corps assets perform GS missions. However, when no corps assets are available and two division platoons are employed as stated above, the two remaining platoons conduct division EPW collection-point operations and other MP functions based on METT-TC. Normally, the EPW platoon and the MP company headquarters colocate in the DSA. As required (and based on METT-TC), airflow planning includes EPW/CI evacuation from the AATF/FOB collection point back to the DSA. The PM section operates from the division rear CP to facilitate I/R operations and to coordinate MMS and AS with key logistical staff. Due to potentially extreme distances on the air-assault battlefield, the DPM normally locates with the division main CP to serve as a key G3 battle-staff member and to coordinate PIO with the G2.

SUPPORT TO DIVISION XXI (HEAVY DIVISION)

7-27. Division XXI represents a significant change in the manner in which division operations are conducted. These changes are brought on by information-age capabilities; an increased integration of service components into an effective battle team; more lethal, survivable, and agile systems; and more capable soldiers and leaders. The Division XXI operates in a larger battle space and at a higher tempo than the AOE division. The division is evolutionary in design, but revolutionary in its use of information technology. It improves the Army's deployability while enhancing its ability to dominate in decisive fights. The following are characteristics of the Division XXI operational environment:

- **Multidimensional.** The division will operate in an extended battle space that goes beyond the traditional physical dimension of width, depth, and height. It includes portions of the electromagnetic spectrum; and it extends beyond the physical boundaries of the division through its communication and digital connectivity to other Army, joint, and coalition elements, even reaching back to CONUS.
- **Precise.** Precision operations go beyond a precision strike to include every aspect of military operations from deployment through combat and through redeployment or transition to other operations. Precision in decisive operations is enabled by three emerging capabilities. First, digitization provides soldiers and leaders at each echelon the information required for making decisions. Second, a full suite of strategic, operational, and tactical sensors linked to analytical teams fuses combat information into situational awareness across the battle space. Third, simulations enable Army elements to be tailored and operations to be planned, war-gamed, and rehearsed—yielding precision execution.

- **Nonlinear.** Nonlinear operations do not seek a battle-space grid of close, deep, and rear operations. Instead, the battle space is fluid, changing as METT-TC changes through the duration of mission preparation and execution. Another dimension of this characteristic is the synchronization of near-simultaneous operations to achieve nonlinear effects across the battle space.
- **Distributed.** Division operations are distributed or executed where and when required to achieve decisive effects concentrated at a decisive point. Dispersion empowers subordinates to operate independently within the commander's intent, leading to synergistic effects that exceed the effects of a centralized headquarters.
- **Simultaneous.** The concept of decentralization operations that are multidimensional, precise, distributed, and nonlinear yields the capability to conduct simultaneous operations across the battle space. Simultaneous operations seize the initiative and present the enemy leadership with multiple crises and no effective responses. Rather than a single, concentrated attack, the division executes a series of attacks (lethal and nonlethal) as simultaneously as possible.
- **Integrated.** Division operations are fully integrated with joint, multinational, and nongovernmental partners. Integrated operations enable the Army to leverage the full suite of capabilities that the services bring to the battle space.

7-28. The Division XXI heavy division is one of total integration—not rounding up or down. Reserve-component soldiers assigned to the division are an integral part of the division mission and task accomplishment. The division cannot operate without them. The division is focused on massing effects on the enemy—enabled by a new foundation of information technology and distributed logistics. The division is built around the pattern of "conduct decisive operations," accomplished by either fires or maneuver. The design performs complementary battle-space shaping in concert with the overall corps plan. Its long-range assets are not typically held in reserve nor just employed in the close fight; they are used to best support the corps and division commanders' plans of operations.

7-29. Although the basic structure of the division has not changed—three maneuver brigades, a division artillery, and a DISCOM—the internal structure and operational construction have changed in a near revolutionary way. The C^2 structure and the basic organizational structure are very similar to the divisions discussed above. However, by integrating the capabilities of digital technology and by using enhanced situational awareness, it is better able to execute combat operations.

7-30. The maneuver brigades have their own scouts—a brigade reconnaissance troop (BRT); therefore, they no longer rely exclusively on the battalion scouts. The maneuver battalions are reduced to three companies with a total number of 45 combat platforms. The division artillery has three Multiple-Launch Rocket System (MLRS) batteries per battalion. The cavalry squadron is under the aviation brigade's C^2 but retains and performs the division's traditional cavalry mission. Figure 7-1 depicts the organizational structure of Division XXI.

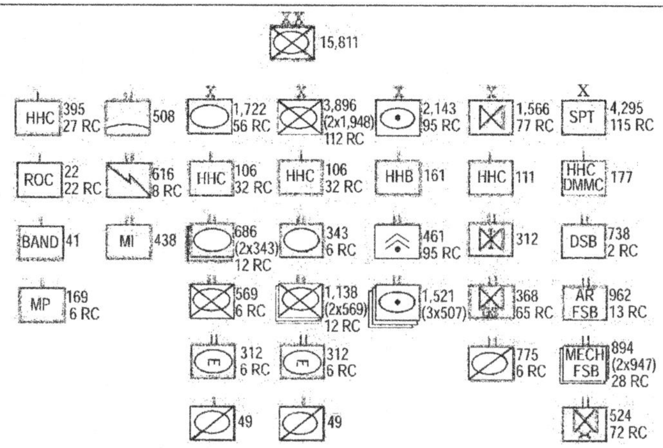

NOTE: Personnel figures are estimates; they may change.

Figure 7-1. Division XXI Structure

7-31. The division's CSS is centralized. Instead of each unit having its own organic CSS personnel, they are now centralized in CSS units. The CSS assets are reorganized and reassigned to the DISCOM. Forward-support companies (FSCs) are habitually associated with each maneuver battalion and forward-support battalions (FSBs) will support each maneuver brigade. This operational structure enables distribution-based rather than supply-based logistics concepts to support the division. The overall logistics footprint for the division is now reduced and the division has organic self-sufficiency to operate unsupported for up to 30 hours.

7-32. A major difference between the AOE fix-forward doctrine and the Division XXI maintenance support is the change in the actual repair site. Division XXI maintenance units directly supporting a given customer will diagnose faults and will replace components forward in the battlefield and will repair rearward. Units will be manned with maintenance personnel capable of performing DS and limited GS maintenance.

7-33. The division operates in a battle space two and a half times larger than an AOE division, and it will usually fight as part of a corps or a JTF. The division conducts offensive, defensive, stability, and support operations in cooperation with other elements of the corps/JTF to gain tactical or operational results significant to the joint-force commander's campaign goals.

7-34. Although the division is a heavy division, MP support is somewhat different from that of an AOE heavy division. The PM (based on METT-TC) establishes operations where MP assets can best be commanded. At the same time, overall MP operations can be synchronized to support operations within the division rear and the division commander's intent can be effectively supported. To this end, the PM conducts split-cell operations within the division main CP. Normally, the PM section is distributed between the security-operations cell (SOC) and the mobility/survivability (M/S) cell. The DPM (who works out of the M/S cell) is also part of the division planning team. He ensures that MP operations are integrated into division plans according to the PM's intent. The DPM attends coordination meetings and assumes coordination responsibility with the rest of the division staff when the PM relocates away from the division main CP.

7-35. The PM, as a staff officer, is involved with every aspect of planning and with the military decision-making process (MDMP). He uses the Force XXI Battle Command—Brigade and Below (FBCB2) C^2 system to understand the battlefield. He understands the mission and the commander's intent and priorities. The PM has complete situational awareness, establishes an effective staff relationship with the G3, and recommends the best way to employ MP resources.

7-36. The Division XXI heavy division MP company has five platoons. Although tasked-organized based on METT-TC and the commander's needs, three MP platoons provide DS for each maneuver brigade. The DS platoon requires a minimum of three squads (each having three teams) to accomplish the mission. The other MP platoons are normally employed forward in sector, providing AS and force protection for the division main and tactical (TAC) CPs and for other HVAs in the vicinity of the maneuver brigade rear boundary. As with the AOE heavy division, the Division XXI heavy division requires augmentation with at least one corps MP company for CS missions in the DRA and can expect two companies in a mature theater (based on METT-TC) as the main effort. In addition, the division may request an MP battalion headquarters to command and control CS MP operations in the rear or to conduct TCF operations.

7-37. While the division MP company is capable of performing all five MP functions, it must be augmented by the corps MP to fully perform all of these functions throughout a mature division's battle space. The PM prioritizes MP functions (based on the supported commander's needs), conducts a mission analysis, and requests needed corps assets through the division G3. Corps MP companies, L&O detachments, and CID assets are needed for sustainment operations. When corps MP companies and L&O detachments are attached to the division, they are placed under OPCON of the PM. The PM will then assign an AO and missions based on METT-TC.

7-38. The PM supports the division's distribution-based logistics concept by conducting aggressive R&S operations that secure the LOC. Since the main effort is to secure the LOC, traditional MMS missions (such as convoy security and escorts) are performed by exception. Through the allocation of increased MP to conduct LOC R&S operations aggressively, the division is able to detect and destroy threat forces before they interrupt operations in rear-area or sustainment operations.

7-39. Digitization gives the division the situational awareness that allows it to cover much larger distances and to move much faster. The systems that provide this digitization are generally unprotected HVAs on the battlefield, which require aggressive AS and force-protection operations.

7-40. The EPW operations are much more fluid, sometimes requiring the division GS MP to move forward and assume brigade collection points, thus allowing DS platoons to move forward with their brigades. Similarly, corps MP may be required to move forward and assume the division's central collection points as the division's rear boundary is pushed forward to shorten the LOC.

7-41. As with all AOE divisions, until corps L&O and CID detachments augment the division, L&O operations are on an exception basis. The MP collect intelligence during every facet of mission execution. The PM develops police intelligence in the division main CP through coordination with the SOC, the G2, the Assistant Chief of Staff, G5 (Civil Affairs) (G5), and the Assistant Chief of Staff, G6 (Communications) (G6).

Chapter 8

Separate Brigades and Initial/Interim Brigade Combat Teams

The MP support separate brigades and IBCTs designed to provide the brigade commander with MP assets that can serve as a force multiplier and that can be employed as an economy of force. The brigade MP platoon is as lethal, flexible, and capable as any other platoon in the MP Corps.

USAMPS

The Army uses separate brigades to inject a small but powerful force where it is needed. It must be able to fight and win while operating on its own for a sustained period of time. It must be able to defend itself on a 360-degree front in war or MOOTW. While there are some variations of separate brigades, it is the heavy separate brigade that is most commonly found within the force.

OVERVIEW

8-1. The Army's IBCT is a full-spectrum, wheeled combat force. It is employed in all operational environments against all projected future threats. However, it is designed and optimized primarily for employment in small-scale contingency operations in complex and urban terrain, confronting low-end and midrange threats that may employ both conventional and asymmetric capabilities. The IBCT deploys very rapidly, executes early entry, and conducts effective combat operations immediately on arrival to prevent, contain, stabilize, or resolve a conflict through shaping and decisive operations. The IBCT participates in war (with augmentation) as a subordinate maneuver component within a division or a corps and in a variety of possible roles. The IBCT also participates (with appropriate augmentation) in stability and support operations as an initial-entry force or as a guarantor to provide security for stability forces by means of its extensive capabilities.

SUPPORT TO SEPARATE BRIGADES

8-2. MP support to separate brigades is provided by a four-squad MP platoon organic to the brigade HHC. A separate PM cell within the brigade HHC serves as the C^2 element for the platoon (see Figure 8-1, page 8-2). Support to the platoon and the PM section for maintenance, supply, mess, and communications is provided by the brigade HHC. Since the platoon and PM section have no organic support, the MP leadership must perform close coordination for this support. However, the MP platoon must compete with other brigade HHC assets for priority of repair for weapons, vehicles, and communications equipment.

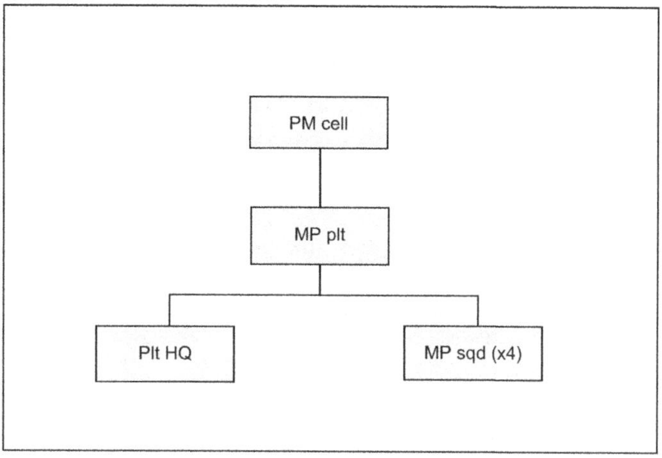

Figure 8-1. MP Support to a Separate Brigade

8-3. The C^2 of an MP unit supporting separate brigades also extends down from the tactical commander. The separate brigade PM has OPCON of brigade MP assets the same way the division PM has OPCON of division MP assets. This includes any MP assets that may have been provided from the corps. The PM advises the commander of a separate brigade on matters pertaining to MP operations. The platoon leader directs the execution of his platoon's mission based on the priorities set forth by the PM and the supported commander. Since the separate brigade's organic MP platoon is more robust than an MP platoon supporting a division maneuver brigade, METT-TC will determine the requirement for augmentation. However, corps L&O and CID augmentation is required.

8-4. The separate brigade's MP platoon is capable of performing all five MP functions. However, its resources are quite limited. Although the MP squads are employed according to METT-TC, the platoon supporting the separate brigade may have—

- One squad operating in the EPW collection point.
- One squad providing a mobile security screen and providing AS around the brigade's main CP.
- Two squads conducting MMS and AS throughout the brigade's rear area.

SUPPORT TO THE INITIAL/INTERIM BRIGADE COMBAT TEAM

8-5. The IBCT is a divisional brigade. It is designed to optimize its organizational effectiveness and to balance the traditional domains of lethality, mobility, and survivability with the domains required for responsiveness, deployability, sustainability, and a reduced in-theater footprint. Its two core qualities are its high mobility (strategic, operational, and tactical) and its ability to achieve decisive action through a dismounted infantry assault. The major fighting components are its motorized infantry battalions. The IBCT has a unique reconnaissance, surveillance, and target acquisition (RSTA) squadron to enhance situational understanding.

8-6. To achieve a rapid deployment threshold, the brigade's design capitalizes on the widespread use of common vehicular platforms—particularly a highly mobile, medium-weight, combat/CS platform coupled with the minimization of the personnel and logistical footprint on theater. Encompassing a personnel strength of about 3,500 and preconfigured in ready-to-fight combined-arms packages, the entire IBCT can deploy within 96 hours of "first aircraft wheels up" and begin operations immediately upon arrival at the APOD. In essence, the APOD is the TAA. The IBCT cannot conduct forced-entry operations, but it provides the JFC with an improved capability to arrive immediately behind forced-entry forces and begin operations to shape the battle space and to execute decisive action to expedite decisions. Once committed, the IBCT can sustain operations for up to 180 days without relief.

8-7. The IBCT's organization is expandable either through augmentation or scalability (according to METT-TC) in any given contingency. The IBCT is scalable in terms of its ability to accept like-type additional forces to expand core tasks and functions already resident within the IBCT (for example, adding additional infantry or RSTA organizations). The IBCT is also capable of accepting augmentation consisting of units or elements that execute tasks or functions not resident within the IBCT (for example, adding armor, air-defense, MP, or aviation assets). In both cases, added units execute their normal mission-essential task list (METL) tasks and, therefore, will not require extensive training to deploy and operate with the IBCT. In both cases, the IBCT includes the command, control, and communications (C^3) capabilities necessary to permit the rapid integration of additional enabling capabilities, particularly for operations outside the scope of small-scale contingency operations (such as stability operations, support operations, and war).

8-8. The IBCT organization excludes other unit-based capabilities often provided in a division slice. However, for each missing capability, the brigade headquarters includes staff cells capable of conducting rudimentary planning and analyses to ensure that all functional-area considerations are incorporated in route planning and preparation for operations. The first MP elements in support of the IBCT are at the brigade headquarters level.

8-9. The MP planning cell is composed of a two-person (a major [MAJ] and a sergeant first class [SFC]) planning cell located within the HHC's maneuver-support cell and under the direct staff supervision of the brigade S3. The role of the MP planning cell is significantly different from that of a traditional division PM or a separate brigade PM. The main difference is in the IBCT's lack of organic or habitual MP assets. However, the absence of organic MP assets makes the job of the MP planning cell that much more critical. The MP planning cell must—

- Understand the organization, capabilities, and limitations of the IBCT.

- Conduct effective liaison with higher headquarters PM elements.
- Become an effective planner and anticipator of MP requirements.
- Prepare MP staff estimates and employment recommendations.
- Plan for MP deployment via air, sea, rail, or land.
- Task-organize MP units effectively and efficiently.
- Assume C^2 of incoming MP forces or, if operating under a division or corps, relinquish C^2 to the division/corps PM or battalion commander (if appropriate for effective employment of MP forces).

8-10. Depending on METT-TC, the brigade may be augmented by MP elements ranging from a platoon to a battalion. In any case, the MP planning cell then becomes a staff planner and coordinator of functional matters pertaining to—

- Force protection, physical security, and vulnerability countermeasures.
- Operations for collecting, processing, and evacuating (possibly) EPWs, CIs, and detainees.
- US military prisoner operations.
- Operations for processing and controlling dislocated civilians and refugees.
- Customs and counterdrug operations.
- Criminal investigations and CID-/MP-investigator support.
- PIO and establishing links with HN police agencies and other international/interagency law-enforcement agencies.
- MMS, AS, and L&O operations.
- Coordination of MP or CID division/echelons above division (EAD) augmentation forces, MWD teams (explosives, narcotics, and patrol), L&O detachments, and MP I/R units through the division, corps, and major Army command (MACOM) PMs.
- Training guidance to other US forces required to perform MP functions in the absence of MP forces.

8-11. Since the IBCT is a divisional brigade, the division PM and the IBCT MP planning cell play an important role in developing an optimum MP force package to support the brigade commander's concept of operations. Despite the brigade's early-entry timelines, the MP planning cell must consider and plan for MP augmentation forces as early as possible to free valuable combat resources. Small-scale contingency operations that result in numerous EPWs, civilian detainees, and refugees will hamper momentum and freedom of maneuver.

8-12. Once the initial brigade receives MP augmentation (see Figure 8-2), the MP priority of effort during the offense will focus on ensuring that routes remain unencumbered and secure for movement of ground combat, CS, and CSS forces. The MP's priority of effort during the defense will focus on conducting AS and counterreconnaissance along the LOC, C^2 centers, and CSS bases. The MP may conduct response-force operations or become part of the TCF.

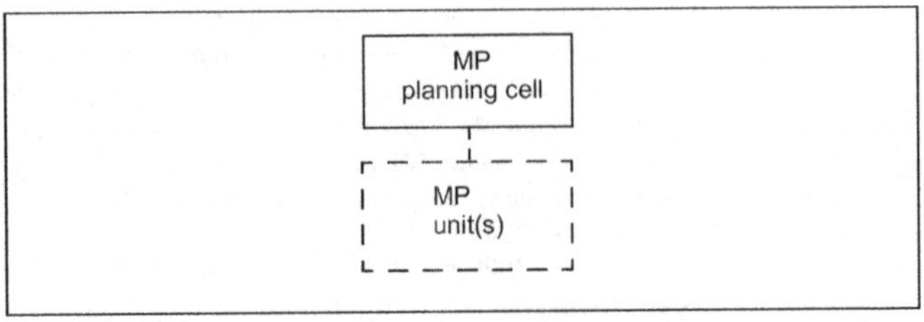

Figure 8-2. MP Support to the IBCT

8-13. Stability and support operations present some unique challenges. The IBCT may be deployed to a geographical area that is politically unstable, that lacks civil control, or that is in complete turmoil. The MP planning cell plans for and coordinates MP support according to available resources and the supported commander's needs. In this scenario (and based on METT-TC), an MP battalion TF could be required to deal with the challenges of—

- A significant number of refugees or dislocated civilians.
- AS or force-protection issues.
- MMS operations.
- Black-market and criminal investigations.
- Restoration of order.
- Civil unrest.
- Intervention of private and nongovernmental organizations.

Chapter 9

The United States Army Criminal Investigation Command

The history of the USACIDC goes back to World War I, when General John J. Pershing organized the CID in France. Today, the USACIDC is the Army's sole agent responsible for investigating felony crimes on and off the battlefield. The USACIDC provides investigative support to commanders at all echelons.

USACIDC

The USACIDC investigates offenses committed against US forces or property, or those committed by military personnel or civilians serving with US forces or where there is a military interest. The USACIDC agents investigate violations of international agreements on land warfare. They conduct special investigations at the direction of the USACIDC's commanding general (CG) or a higher authority. In general, the USACIDC's missions include—

- Investigating and deterring serious crimes.
- Conducting sensitive/serious investigations.
- Collecting, analyzing, processing, and disseminating criminal intelligence (CRIMINTEL).
- Conducting protective-service operations for designated personnel.
- Providing forensic-laboratory support.
- Maintaining Army criminal records.
- Enhancing the commander's crime-prevention and force-protection programs.
- Performing LOGSEC operations.

OVERVIEW

9-1. The USACIDC's operations help the commander maintain discipline and order by preventing or investigating crimes that reduce a unit's ability to fight. During the investigation of serious crimes, the USACIDC concentrates its efforts on investigating serious crimes such as wrongful deaths, controlled-substance offenses, theft, fraud, sex crimes, and assaults. The USACIDC also conducts sensitive and special investigations involving matters pertaining to senior Army officials and those associated with classified programs.

9-2. The USACIDC agents collect, analyze, process, and disseminate criminal intelligence/information relating to crime within or directed toward the Army. Specific information relating to modus operandi, crime techniques, investigative leads, gang violence, and terrorism is shared with the appropriate intelligence and law-enforcement agencies. Conversely, USACIDC agents solicit and receive crime-related information from the MP and from local, national, and foreign law-enforcement and intelligence agencies. Special agents identify and evaluate crime-conducive conditions and indicators of potential attacks against Army property, facilities, or personnel. They then provide reports to the appropriate commander.

COMMAND AND CONTROL

9-3. The USACIDC is a centralized (stovepipe) MACOM whose special agents in the field report through the USACIDC's chain of command (detachment to battalion to group) to the CG, who in turn reports directly to the Army Chief of Staff and the Secretary of the Army (see Figure 9-1).

9-4. As an independent MACOM, the USACIDC's tactical units are not under the C^2 of supported organizations. The USACIDC's elements may be temporarily attached to a supported commander when required to accomplish a nonstandard mission. Reassignment, promotion, accreditation, and disciplinary actions are retained by the USACIDC. Attachments will be coordinated with the appropriate USACIDC headquarters planning the specific mission and approved by the CG, USACIDC. Additionally, although there is no formal staff relationship, USACIDC commanders advise their supported commanders on criminal-investigation matters. This enhances the quality, reliability of information, support, and trust between USACIDC elements and their supported commanders.

WARTIME SUPPORT

9-5. The USACIDC supports each echelon of command from the division to the ASCC. The theater USACIDC structure is comprised of a C^2 headquarters

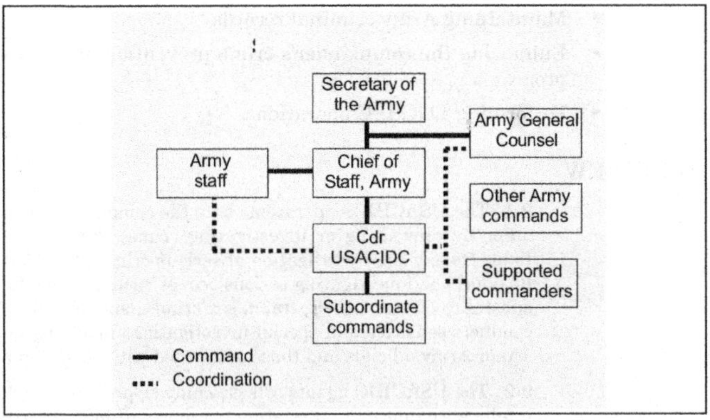

Figure 9-1. USACIDC Chain of Command

and mobile, modular, and tailorable investigative detachments. The USACIDC supports combatant commanders with the following functions:

- **LOGSEC.** Tracking and protecting materials and equipment from the manufacturer to the soldier on the battlefield.
- **CRIMINTEL.** Collecting, consolidating, analyzing, and disseminating intelligence associated with criminal and terrorists activities targeted at Army interests.
- **Criminal investigations.** Investigating war crimes and, in some cases, crimes against coalition forces and HN personnel.
- **Protective-service operations.** As in peacetime, protecting key personnel anywhere on the battlefield.

9-6. The USACIDC's LOGSEC function protects the Army's supply pipeline against criminal activities from the manufacturer, through logistics channels, all the way to the frontline soldier. It involves preventing, detecting, and investigating criminal and terrorist activities such as supply diversion, destruction, and sabotage or product substitution. The USACIDC's LOGSEC operations assess LOGSEC, identify weaknesses, and provide a prioritization of threats so that commanders can implement preventive measures to reduce the vulnerability of the logistics pipeline. Whenever possible, the USACIDC will initiate actions to recover logistical losses and return them to Army control.

The Army let a $22 million contract for the manufacture of its mobile kitchen trailer (MKT)-90 mobile field kitchens in 1990. These kitchens were distributed to Army units worldwide, including those units serving in Saudi Arabia during Operation Desert Shield/Desert Storm. After distribution of the MKT was well underway, a US government quality-assurance representative received information from the manufacturer's employees indicating that the kitchens were not built to contract specifications. This information led to an investigation by the USACIDC's major fraud procurement unit (MFPU). The investigation revealed that inferior material and manufacturing processes were substituted, which allowed the contractor to realize an additional $228,000 in profit. The US Army Environmental Hygiene Agency determined that the MKT-90 had serious sanitation defects and posed a safety hazard to troops using and being fed from these kitchens. The problem with the MKT was identified and expeditiously resolved to ensure the safety of soldiers in the field. The primary subject was indicted by a grand jury and subsequently pled guilty to one count of major fraud. He was found guilty, sentenced to 10 months confinement, and ordered to pay restitution.

9-7. The USACIDC's CRIMINTEL collection efforts focus on the identification and prevention of terrorist and nonterrorist crimes against US and allied military personnel, facilities, and other interests. Upon collecting this information, agents recommend countermeasures to combat subversive activities through coordination with MI, the PM, rear-area operations officers, and HN military and civil intelligence agencies. Additionally, when directed, the USACIDC becomes the lead US military investigative agency at theater level. When this happens, it is tasked with leading the prevention-of-terrorism effort from all services, not just the Army component.

> The MI analysts, the MP, and US Army criminal investigators deployed in Kosovo formed a combined all-source information center. This center provided threat assessments to units assigned to TF Falcon. Fifty-five percent of these assessments referenced some form of criminal activity. These intelligence products provided commanders with reliable information concerning potential incidents or criminal activity. With nearly 7,000 US soldiers deployed within the region, these assessments were absolutely crucial for the force-protection planning effort.
>
> The fusion cell within the information center developed intelligence products from national, theater, and operational sources. Due to the significant threat posed by criminal elements in the region, two CID military agents and two CID civilian analysts were attached to the fusion cell to facilitate the police-intelligence function. The CID personnel, in cooperation with MP soldiers, played a key role by linking CRIMINTEL to specific groups and events. The CRIMINTEL collection effort was specifically targeted on weapons, drugs, organized crime, and the identification of smuggling routes. The identification of smuggling routes resulted in a significant increase in the number of weapons being confiscated. The timely transfer of CRIMINTEL products to tactical units enabled a rapid response to serious confrontations, increased confiscation of arms and ammunition, and improved stability in TF Falcon's AO.

9-8. During war and MOOTW, the criminal-investigation effort includes—

- War crimes and, when directed, crimes against coalition forces and HN personnel.
- Major felony crimes committed by US personnel or EPWs/CIs and, when directed, crimes committed by foreign nationals in which there is an Army interest.
- Criminal acts by indigenous personnel, factions, and ad hoc groups.
- Special/sensitive investigations, hostage negotiations, polygraph operations, and force-protection operations.
- CRIMINTEL operations.

9-9. The USACIDC detachments provide enhanced coverage and protection for designated key and essential leaders during war and MOOTW. The protection requirement for senior JTFs and Army commanders may be significantly greater during MOOTW than during peacetime or war as the propensity of asymmetrical threats (such as criminal and terrorist groups) operating in the AO also increases.

ORGANIZATION

9-10. The USACIDC is organized with an array of multifunctional units. As with many other Army organizations, its force is in a transition process. During peacetime, the USACIDC groups provide brigade-level area support for worldwide Army operations. The group provides C^2, staff planning, and administrative support for USACIDC elements assigned to an AOR or deployed to a contingency operation. During war, a group provides the same capabilities in support of an ASCC and provides C^2 for all USACIDC units in the theater. Regardless of the operational environment, the group ensures the connectivity between all USACIDC units. It establishes and maintains links with supported units and interagency, joint, multinational, allied, and HN authorities on matters pertaining to Army and USACIDC operations. The group headquarters has a command section, a detachment headquarters, an S1, an S2, an S3, an S4, an SJA, and a communications section. It provides C^2 for up to six subordinate battalions. Major differences in the AOE and Force XXI groups are in an increase of support and mission personnel and the movement of all polygraph support down to battalion level.

9-11. During peacetime, the USACIDC battalions provide area support for worldwide Army operations. They provide C^2, staff planning, forensic-science support, CRIMINTEL, polygraph support, technical guidance and supervision of USACIDC operations, and administrative support for USACIDC elements assigned to an AOR or deployed to a contingency operation. During war, the USACIDC battalion provides the same capabilities in support of each corps and TSC. Under AOE designs, one battalion supports each corps and TSC. Under the Force XXI design, up to two battalions support each TSC. Like the groups, each USACIDC battalion ensures connectivity and establishes and maintains links with other units/agencies within its AOR. The battalion headquarters consists of a command section, a detachment headquarters, an S1, an S2, an S3, an S4, and a communications section. It provides C^2 to five USACIDC detachments on a sustained basis. Its span of control can accommodate up to seven detachments for a short period of time. Major differences in the new battalion are the addition of a CSM, an S6 officer, a detachment commander and support personnel, and CRIMINTEL and forensic personnel.

9-12. The USACIDC investigative team is the smallest operational investigative element. Each team consists of two special agents (a warrant officer and a noncommissioned officer [NCO]). METT-TC requirements dictate that these teams have the capability to operate independently from the detachment headquarters. The investigative teams are the building block for both AOE and Force XXI USACIDC detachments. Both designs provide levels of flexibility to task-organize without splitting units apart. The detachment (with its complement of teams) performs the full range of criminal-investigative functions in all operational environments, but it has limited CRIMINTEL and LOGSEC management capabilities and no forensic or polygraph capability. Each detachment includes a headquarters section and up to eight investigative teams. METT-TC requirements dictate that these teams have the capability to operate independently from the detachment headquarters. The major differences between the AOE and Force XXI detachments' design begin with replacing the heavy and light division support elements (DSEs), the corps-area support element (CASE), the TSC-area support element, (TASE), and the port-area support element (PASE) designs with a modular design consisting of a standardized headquarters element and two special-agent sections. Each section consists of four investigative teams. Personnel administration, supply, and the bulk of administrative support found in the AOE organizations have been consolidated at battalion.

Additionally, the Force XXI detachment is commanded by a special-agent warrant officer.

9-13. Both the AOE and the Force XXI units retain the Army's standard dependencies upon other units for support requirements (religious, health services, finance, photographic processing, and so forth). The new USACIDC TOEs have enhanced A/L support capabilities. However, the new units still retain the AOE support requirement for food services and emergency Class III/V resupply from CS MP brigades, battalions, and companies.

THE FIELD INVESTIGATIVE UNIT

9-14. Sensitive investigations are normally conducted by the field investigative unit (FIU). The FIU is a one-of-a-kind organization within the DOD that enables the Secretary of the Army to conduct sensitive investigations requiring access to special information or programs that are highly classified. The FIU works closely with the Army IG, the Judge Advocate General of the Army, and the Army General Counsel to support commanders in the special-operations and intelligence communities and in the area of classified acquisition programs. The FIU may also be assigned investigations involving senior Army personnel or those of special interest to the Army leadership.

THE COMPUTER-CRIME INVESTIGATION UNIT

9-15. The computer-crime investigative unit (CCIU) deals with intrusions involving classified networks or multijurisdictional offenses. The CCIU works closely with MI and federal law-enforcement agencies to coordinate military actions, nonmilitary-affiliated offenders, and foreign-intelligence services. In addition, it provides technical assistance to CID elements that are conducting computer-related investigations.

THE PROTECTIVE-SERVICE UNIT

9-16. The protective-service unit (PSU) provides worldwide protective services to designated personnel to protect them from assassination, kidnapping, injury, or embarrassment. The PSU plans, coordinates, and executes executive protection for—

- The Secretary of Defense.
- The Deputy Secretary of Defense.
- The Chairman of the Joint Chiefs of Staff.
- The Vice Chairman of the Joint Chiefs of Staff.
- The Secretary of the Army.
- The Army Chief of Staff.
- The Army Vice Chief of Staff.
- Others (as directed).

9-17. During war and MOOTW, the USACIDC routinely provides protective-service support to senior Army ground commanders and, in some cases, to JTF commanders. The USACIDC agents also provide training to the MP details providing protective services to senior field commanders.

The President of the US hosted the North Atlantic Treaty Organization's (NATO's) 50th summit anniversary celebration in Washington, DC during the period of 22 to 26 April 1999. The President invited 28 heads of state or prime ministers, 48 foreign ministers, 88 ministers and chiefs of defense, and the chairman of the military committee to the summit. During the conference, the principal guests resided in 23 hotels in Washington, DC. The responsibility for providing protection for the ministers, chiefs of defense, and the chairman for the military committee was given to the USACIDC's PSU. In preparation for this mission, additional training on protective-service operations and evasive driving techniques was given to over 200 reserve agents. These individual mobilization augmentees, troop program units, and other USACIDC agents were critical to the successful execution of this mission.

US ARMY CRIMINAL-INVESTIGATIONS LABORATORY

9-18. Forensic-laboratory support is provided by the US Army Criminal-Investigations Laboratory (USACIL) that is currently located at Fort Gillem, Georgia. The USACIL is a nationally accredited forensic laboratory that provides all three services with dedicated, highly responsive, deployable, state-of-the-art scientific and forensic investigative support. The six forensic divisions of the USACIL include the following:

- Chemistry (serology, drugs, trace, and deoxyribonucleic-acid [DNA] testing).
- Questioned documents (alterations and comparisons of handwriting, typing, and printing).
- Firearms and tool marks (weapons and obliterated numbers).
- Photography (evidence, charts and photos for court).
- Latent prints.
- Computer forensics.

9-19. The USACIL collects, receives, and examines evidence and prepares reports of findings. It also provides expert-witness testimony at court-martials.

US CRIME RECORDS CENTER

9-20. The US Crime Records Center (CRC) maintains the Army's criminal records. It maintains more than two million MP reports and USACIDC reports of investigations. These records are retained for at least 40 years before being destroyed. During their retention, the records are actively maintained and frequently researched.

During the course of an investigative search, civilian police authorities seized 19 hand grenades. A routine search of stolen property and criminal index files failed to identify the grenades as having been stolen from a nearby military installation. However, a search of the CRC files identified a USACIDC investigation relating to the theft of a case of grenades that occurred 20 years earlier. Further investigation revealed that the lot number of the recovered grenades was identical to that of the stolen grenades. The investigative leads resulting from the CRC file search contributed to the resolution of both cases.

9-21. The CRC conducts more than 10,000 criminal-history name checks each month to identify victims and perpetrators of criminal offenses. The checks are requested not only by USACIDC agents, but also by other military and civilian law-enforcement officials. The CRC is also the Army's agent for Freedom of Information Act (FOIA) and Privacy Act requests relating to criminal investigations and MP reports. The CRC also manages the Army's polygraph programs and their support to Army installations around the world.

EMPLOYMENT

9-22. Whenever possible, USACIDC units are colocated with the MP. This provides unified MP support across the battlefield, and it facilitates logistical support (primarily mess) provided by MP units. However, the USACIDC maintains its vertical C^2. Based on METT-TC, the MP group (CID) will colocate with the supported ASCC and the MP battalions (CID) supporting the corps and EAC will colocate with the MP brigade (CS). When feasible, MP detachments (CID) will colocate with MP battalions/companies and division MP companies. Figure 9-2 illustrates a typical battlefield array for AOE organizations, and Figure 9-3 illustrates a Force XXI array.

9-23. Since USACIDC detachments are austere organizations with little organic A/L capability, they rely on the battalion for unit-level maintenance, supply, and personnel support. However, detachments must often operate independently when performing criminal-investigation functions throughout the AOR. During these times of long-distance separation, detachments must seek A/L support from the supported unit.

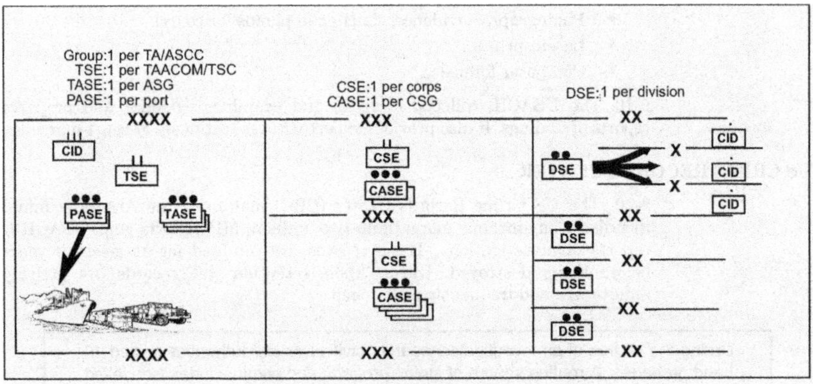

Figure 9-2. AOE Battlefield Array

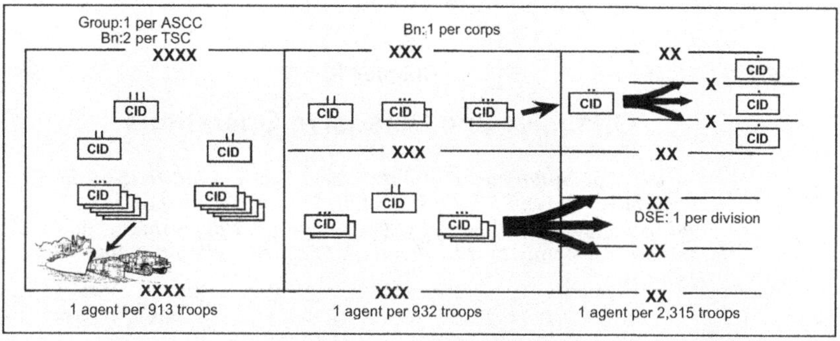

Figure 9-3. Force XXI Battlefield Array

Chapter 10

Offensive and Defensive Operations

The whole of military activity must relate directly or indirectly to the engagement. The end for which a soldier is recruited, clothed, armed, and trained; the whole object of his sleeping, eating, drinking, and marching is simply that he should fight at the right place and the right time.

Clausewitz

Army doctrine considers the full range of operations from war to MOOTW. An operation is a military action or the carrying out of a military mission needed to gain the objectives of any battle or campaign. Offensive and defensive operations normally dominate military operations in war and in some smaller-scale contingencies. An offensive operation is aimed at destroying or defeating the enemy and imposing our will on him for a decisive victory. Defensive operations defeat an enemy attack, buy time, economize forces, or develop conditions favorable for a counteroffensive that regains the initiative and defeats the enemy.

OVERVIEW

10-1. The offense is the decisive form of battle. Offensive campaigns and major operations are designed to achieve operational and strategic objectives quickly and decisively at the least cost. Offensive operations combine the terrain and the force to achieve the objective. The main purpose of the offense is to defeat, destroy, or neutralize the enemy force. Offensive operations are characterized by a rapid shift in the main effort to take advantage of opportunities by momentum and simultaneous destruction of enemy defenses.

10-2. The immediate purpose of defensive operations is to defeat an attack. Military forces defend only until they can gain sufficient strength to attack. Commanders choose to defend when they need to buy time, to hold a piece of terrain, to facilitate other operations, or to erode enemy resources at a rapid rate while reinforcing friendly operations. The ARFOR conduct defensive operations as part of major operations and campaigns and in combination with offensive operations (see FM 3-0).

SUPPORT TO OFFENSIVE OPERATIONS

10-3. MP support to offensive operations varies according to the type of operation being conducted. The MP functions support each echelon commander based on METT-TC. In the offense, MP priorities are often placed on MMS, I/R, and AS operations. Regardless of the offensive effort, MP leaders supporting an offensive operation must—

- Understand the commander's intent and his concept of operations.
- Anticipate that selective elements of the offensive force may need to pause, defend, resupply, or reconstitute while other forces attack.
- Anticipate changes in the operational tempo and prepare to weight the MP effort to meet the maneuver commander's intent.
- Provide MP support within the context of linear and nonlinear operations. They must understand how both types of operations affect the security of rear areas and LOC, the protection of C4I, the means of sustainment, and the land force's mobility.
- Support the maneuver commander's intelligence efforts by conducting the PIO function. The MP must observe, identify, and report information.
- Assist maneuver forces to mass. The MP protect C^2 nodes such as the CP that orchestrates the mass efforts and resources that sustain the offensive move. The MP must deny incursions by enemy reconnaissance forces into the AO. They also maintain surveillance, provide early warning, and impede and harass the enemy with supporting and organic fires.
- Speed the commitment of follow-on forces and reserves. They must also conduct MMS operations to assist the commander in a smooth and quick shift of forces to support the main effort.
- Anticipate the pursuit and exploitation by positioning MP forces that can support continuous operations. The MP support as far forward as possible while protecting resources needed to seize the opportunity. The EPWs/CIs are collected as far forward as possible. The MP coordinate the treatment of dislocated civilians and refugees with the HN or foreign forces.
- Tailor the MP support based on the unit being supported (corps, divisions, and brigades).
- Consider the location and composition of response forces and the TCF and the strength of rear-area threats.
- Integrate a force into the offensive plan that is capable of conducting aggressive combat operations against designated targets.

> During Operation Just Cause, an MP platoon received the mission to disable an armed Panamanian Defense Force (PDF) patrol boat docked near a key objective. After receiving the mission, the platoon leader maneuvered his force into position and attacked after receiving enemy fire. The platoon successfully disabled the patrol boat and captured the objective while suffering minimum casualties.

10-4. During offensive operations, MP support provided by teams, squads, and platoons provide the commander with an agile, flexible, versatile, and capable force ready to contribute to the overall mission success.

SUPPORT TO DEFENSIVE OPERATIONS

10-5. The MP support defensive operations consistent with the commander's intent and priorities. They support a defending force by assisting subordinate elements in the execution of different, yet complementary actions. For example, the MP may support the maneuver of an attacking unit by conducting MMS operations while simultaneously providing AS for a defending unit in a base cluster. The MP support reflects the maneuver commander's focus—destroying an attacking force, retaining or denying key terrain, moving away from an enemy force, or a combination of the aforementioned. However, despite the intent of the defensive operation being conducted, MP leaders must—

- Conduct MMS operations to help a force maneuver and mass. The MP must anticipate the shift from the defense to the offense and assist the forward movement of reserves.
- Conduct AS operations to deny information to enemy reconnaissance elements seeking out the exact location of the defending force. The MP are positioned where they can control key terrain or improve the defensive capability of bases and base clusters. The MP conduct aggressive R&S to locate and deny enemy access to critical logistical facilities.
- Protect sustainment resources while supporting the lateral, forward, and rearward movement of forces.
- Consider the type and size of the AO, the LOC security, the rear threat, and the number of EPWs/CIs and dislocated civilians to determine how they may affect the movement of forces.

> During Operation Desert Shield, EAC, corps, and division MP units conducted a myriad of missions to support both the XVIII Corps's (Airborne) and the VII Corps's movement to their TAA. The MP's main focus was on MMS and AS throughout the AO to ensure freedom of movement and to protect US and coalition forces against terrorists and other threats in the rear area.

10-6. The success of MP operations in the defense depends on the MP leader's situational awareness, the commander's intent, and the precise employment of MP resources when and where they are needed. In the defense, the MP perform their five functions to assist the defending commander by providing a lethal mobile force that allows him to concentrate his efforts and resources in fighting the next battle.

Chapter 11

Stability and Support Operations

ARFOR conduct stability operations in a complex, dynamic, and often asymmetric environment. Stability operations are usually nonlinear and noncontiguous. They are often time and manpower intensive. The purpose of support operations is to meet the immediate needs of designated groups for a limited time until civil authorities can accomplish these tasks without Army assistance.

FM 3-0

The Army's mission is to prepare for war and, when deterrence fails, to achieve a quick, decisive victory. The DOD projects that many of its future missions will be stability and support operations. However, these operations are not new to the Army. The Army has participated in stability and support operations in support of national interests throughout history. The Army has protected its citizens, assisted nations abroad, and served America in a variety of other missions. The pace, frequency, and variety of stability and support operations have quickened in the last three decades. As demonstrated in Operations Provide Comfort, Restore Democracy, and JTF Los Angeles, these types of operations present a challenge to all services.

OVERVIEW

11-1. Stability operations are conducted outside of the US and its territories to promote and protect US national interests by influencing political, civil, and military environments and by disrupting specific illegal activities. Stability operations may include both developmental and coercive actions. Developmental actions enhance a HN government's willingness and ability to care for its people. Coercive actions apply carefully prescribed limited force or the threat of forces to change the AO's environment.

11-2. Support operations provide essential supplies and services to relieve suffering and to help civil authorities prepare or respond to crises. In most cases, the ARFOR's efforts are focused on overcoming conditions created by man-made or natural disasters. The ultimate goal of support operations is to meet the immediate needs of designated groups (for a limited time) until civil authorities can accomplish these tasks without Army assistance.

11-3. Stability and support operations are distinctly different from offensive and defensive operations and must be analyzed differently. For instance, the application of the rules of engagement (ROE) will be unique to the operation. The restrictions on combat operations and the use of force must be understood and obeyed at all levels. MP leaders must clearly take the initiative and ensure that the applicable guidance is implemented and followed by all subordinates.

11-4. Stability and support operations have more diverse political considerations than are usually encountered in war. There are also complexities that must be addressed in joint, combined, and interagency stability and support operations. The Army's challenge is to execute these operations successfully in a joint or combined arena without degrading its capability to wage war successfully.

11-5. If the US is to fight and win future wars, it must develop, train, and employ its force structure in a manner that ensures success in every operation across the spectrum of military operations. Under these auspices, MP forces can provide commanders with especially valuable assets for stability and support operations. The MP have the unique capability of serving as a combat, CS, and CSS force. The domestic and international acceptability of the MP operational image frequently makes the MP the most appropriate force for stability and support operations. Additionally, the MP are flexible and capable of rapidly transitioning from one end of the spectrum to the other if the environment changes.

FORCE SUITABILITY

11-6. The MP's capabilities and their operational and supporting tasks are as integral to stability and support operations as they are to offensive and defensive operations. MP branch-/core-related skills are highly compatible with the capabilities required for stability and support operations. The MP provide a highly capable, politically acceptable force that is suitable for a variety of missions. They possess robust moving, shooting, and communicating capabilities; and they project an assist, protect, and defend image that is particularly important when tailoring a force that requires significant capabilities but a low political profile. The MP provide a flexible, but limited, economy-of-force organization. Mission requirements will always exceed available MP resources, especially in stability and support operations. For this reason, MP assets must be prioritized to deliver the greatest mission capability. The specific operations that MP units perform at a given time are determined by the echelon commander's needs, the mission requirements, and the availability of MP resources. As discussed in previous chapters, the echelon commander, through the command's PM, sets the priorities for MP operations. As in offensive and defensive operations, the PM knows that while the MP force can perform all MP functions, they may not be robust enough to perform all assigned MP missions all of the time. The PM must continuously make trade-offs between the numbers and kinds of MP missions that can be supported.

11-7. The MP's capabilities are further enhanced by their training, mind-set, and experience in dealing with people in highly stressful and confusing situations. Although capable of conducting combat operations when needed, the MP are highly practiced in de-escalation and in employing the minimum-essential force to contain potentially violent situations. This mind-set serves as the framework for MP law-enforcement training and is especially applicable in MP support for stability and support operations. Furthermore, this mind-set is exercised and reinforced daily in peacetime law-enforcement operations. This is a significant capability. MP soldiers learn and receive constant reinforcement training in controlling a situation on the spot before it escalates and cannot be controlled. The MP have a unique understanding of human nature and are adept at handling emotion-laden situations.

> During operations in Kosovo, as both Serbs and Albanians attempted to create unfavorable conditions and responses by US forces, the MP kept their cool under fire in the streets of Gnjilane and Kamenica. The MP's reaction led to a quick understanding by all sides that the US Army MP were fair and just and understood the tense situations.

11-8. The MP stress four essential training competencies—skill proficiency, human dimension and attitude, camaraderie and teamwork, and leadership. These competencies are the foundation for all MP operational performances. They equip the MP to exercise discretion in dealing with others, to protect and assist those in need, and to accomplish the varying demands of MP missions in stability and support operations and, when needed, in war.

STABILITY AND SUPPORT

11-9. Stability and support operations are performed in support of US authorities or in support of international bodies and foreign governments when so required by US authorities. The MP have long provided the essential support for stability and support operations; however, the frequency of these operations has increased significantly in recent years. Consequently, the MP have been deployed in greater numbers with greater frequency. MP units have a substantial history of successful, simultaneous, and continuous deployments in both a lead and a support role.

SUPPORT TO DOMESTIC CIVIL AUTHORITY

11-10. The purpose of support to domestic civil authorities is to—

- Render humanitarian assistance.
- Provide disaster relief.
- Restore order.
- Combat terrorism.
- Support counterdrug operations.

11-11. Military operations supporting domestic civil authorities are governed by the provisions of the Stafford Act, the Posse Comitatus Act, and other laws and regulations. Peacetime experience in civil-military affairs equips the MP to oversee, assist, and train other units and agencies for law enforcement, sensitive security, and operations to control dislocated civilians (within the legal limits of a particular stability or support operation). For this reason, the MP are often needed after other forces have redeployed, and they stay until government order is restored. Since active-component military missions are constrained by the Posse Comitatus Act, nonfederalized national guard (NG) units are often employed to conduct these operations (particularly counterdrug operations). The NG units can operate under gubernatorial authority within state jurisdiction without Posse Comitatus restrictions. The NG MP units are the only NG units trained, skilled, and experienced in providing law enforcement. They offer civilian authorities a unique, specialized capability not available from other NG units.

SUPPORT TO HUMANITARIAN ASSISTANCE AND DISASTER RELIEF

11-12. Humanitarian assistance and disaster-relief operations provide emergency assistance to victims of natural and man-made disasters and other endemic conditions. Natural disasters include earthquakes, floods, and hurricanes. Man-made disasters include riots and civil conflicts. Humanitarian assistance and disaster relief conducted by US forces alleviate urgent nonmilitary needs of a populace until the appropriate civilian agencies can provide the needed services without military support. These types of operations include, but are not limited to—

- Conducting searches and rescues.
- Evacuating, sheltering, sustaining, and protecting dislocated civilians.
- Making food, medical care and treatment, and other civilian-aid programs available.
- Reducing opportunities for criminal activity and restoring L&O.
- Maintaining other operations needed to ensure the well being of the affected population.
- Providing relief operations for man-made disasters. These operations are characterized by civil conflict, and they entail all of the requirements for other disasters but focus on the restoration of L&O.
- Providing force protection and security.

11-13. MP support can be an important asset in all disaster-relief operations. The MP are trained and equipped for decentralized operations. MP teams operate in highly mobile vehicles equipped with radios, which are invaluable in disaster-relief operations. Each MP platoon, company, and battalion headquarters is equipped with high-frequency radios. These capabilities allow the MP (even when widely dispersed) to maintain centralized communications; to provide L&O support; and to assist in search, rescue, and evacuation efforts. Because of their decentralized operations and communications equipment, the MP can also play a vital role in notification, ADC, and access and egress control. They can deploy with disaster-assessment teams to assist in advising and developing a disaster-assistance plan. The MP can also deploy to an affected area to—

- Assist in restoring order.
- Search for, rescue, and evacuate victims.
- Provide personnel and vehicular circulation controls.
- Provide ADC.
- Conduct detention operations.
- Provide physical and area security to protect life and property immediately following a disaster.
- Protect fire and emergency-services personnel.
- Gather, record, and report information.
- Provide force protection and security.

11-14. MP (I/R) battalions can support civil-affairs units by establishing I/R facilities to shelter, sustain, protect, and account for disaster-relief victims. These units can also train and assist government agencies (US and foreign) or police in the operation of facilities for dislocated civilians and detainees.

11-15. The MP possess capabilities critical to relief in man-made disasters characterized by civil conflict. The MP have the essential training and skills needed for relief operations that focus on restoring civil order. Restoration of civil order demands—

- **Objectivity and neutrality.** The objective of civil-disturbance control is to restore L&O, not to impose punishment or suppress peaceful dissent. Government forces quelling civil disorders must be emotionally objective and politically neutral. MP law-enforcement training stresses objectivity and neutrality, an important attribute for forces employed in stability or support operations.
- **Minimal use of force.** The application of more than minimum force may be detrimental to mission success. The MP are particularly suited for these situations, having been trained on the technique for the use and measured application of minimum forces to accomplish mission goals.
- **Public acceptance.** Government forces must avoid actions that will alienate the populace. Any action offensive to community sensitivities must be avoided. The MP receive training in crowd-control operations, interpersonal communications skills, and minimum use-of-force techniques—skills that are critical in gaining public acceptance.
- **Positive image.** In many disturbance operations, demonstrators and government forces compete for the approval and sympathy of the general population. The MP's assist-and-protect image provides the government with a distinct advantage in such situations.
- **Threat awareness.** Government forces must pay attention to and try to learn the organization, intentions, and activities of demonstrators. Equal attention must be paid to how the community will respond to the demonstrator's actions and the government's response. The MP interact daily with the populace, placing them in a position to develop criminal information and to determine the collective attitude of the populace.
- **Effective coordination.** Disturbance operations often involve many agencies from various levels of government. Thorough, timely, and coordinated preparation, planning, training, and execution are essential for mission success. MP civil-disturbance training (including planning and coordination measures) and their communications assets enhance coordination, particularly during mission execution.
- **Operational unity.** Multiple control forces representing different agencies with varying jurisdictional authority can pose C^2 problems. When unity of command cannot be achieved, operational unity must be achieved by colocating operational centers, integrating communications, and delineating organizational responsibilities. The MP are well trained in police and military operations, possessing a sound understanding of both civilian police and military measures, legal constraints, and effective coordination.

11-16. The MP operational tasks supporting man-made disasters vary according to the nature of the disaster and the operational environment. For example, in CONUS where the commitment of military forces to civil disturbances is considered a last resort and military involvement is limited by law, nonfederalized NG MP units have long been employed by civil authorities for their unique capabilities. MP forces can—

- Isolate and contain an area with barricades, roadblocks, and perimeter patrols. The MP can also assist civil-affairs units by enforcing pass and identification systems and public-utility controls.
- Provide mobile patrols to enhance the security of high-priority targets (such as buildings, utilities, and services that are critical to the community's economic and physical well being).
- Monitor, disperse, contain, or limit crowd movement through—
  - Observation points and patrols.
  - The communication of interest and intent to the crowd.
  - The channelization or diversion of the crowd.
  - Gaining the cooperation of the dissident leadership.
  - The proclamation of the illegal nature of the crowd's actions.
  - Show of force.
  - The use of crowd-control formations.
  - The apprehension of crowd members and leaders.
  - The application of minimum-force measures.
- Establish area control to prevent looting, to protect businesses and other likely targets, and to prevent arson. The MP accomplish this mission with saturation patrolling, including vehicle and foot patrols. With augmentation, the MP can perform air and water patrols. They also enforce populace control measures (such as ordinances to prevent gathering, permits to gather, restrictions on circulation, restrictions on the interference with government and public functions and personnel, restrictions on possessing weapons, and other measures instituted by civil authorities).
- Neutralize special threats (such as snipers and bombings) that are highly dangerous to both government forces and the community. Snipers pose a particularly grave danger in disaster-relief operations. The MP can take immediate protective actions, secure an area, isolate the threat and, when authorized, use SRTs to assist civilian or HN authorities in apprehending or neutralizing the sniper. The MP are trained to perform the necessary actions when an explosive device is discovered. In such an event, they can secure and evacuate an area, organize search teams, and isolate the site.

11-17. The MP can also support in rescue and recovery operations, which may include rescuing US or foreign nationals. Such operations are also employed to locate, identify, and recover sensitive materials deemed critical to national security. The operations may be conducted in benign or hostile environments. The MP provide OPSEC, and they process and account for civilians affected by rescue and recovery operations. The MWD teams can also help search for lost or hidden personnel. In stability and support operations, the MP can perform their police functions for rescue and recovery operations while keeping the perceived military signature low.

SUPPORT TO COUNTERDRUG OPERATIONS

11-18. Counterdrug operations are measures taken to disrupt, interdict, and destroy illicit drug activities. In a counterdrug role, the DOD may offer certain forms of support to HN counterdrug personnel, to US civilian law-enforcement agencies operating in a HN, to Department of State (DOS) counterdrug personnel, and to the Bureau of International Narcotics Matters. Without direction from the National Command Authority acting under constitutional or statutory authority, US forces engaged in counterdrug activities may not engage in direct law-enforcement activities (such as apprehension, search, seizure, and other similar activities). As discussed earlier, the Posse Comitatus Act authorizes specific DOD assistance in counterdrug activities, but US military forces usually may not participate directly in arrests, searches, seizures, and similar activities. Conversely, as discussed earlier, USC Title 32 authorizes state governors to use NG MP units with full authority to engage directly in law-enforcement activities. When authorized, US advisors may accompany counterdrug forces on operations.

11-19. MP operational contributions can include the use of force, drug identification, crime-scene protection, customs operations, surveillance of criminal activity, serious-incident reporting, and other related activities. Employing the MP for these operations ensures force suitability and avoids a mismatch in terms of unit capability and mission effectiveness.

11-20. Border-screening operations consist of three separate but related measures—mobile patrolling, the use of listening/observation posts, and airfield surveillance. These measures are part of the MP's AS and MMS functions used in tactical rear-area operations. The MP's operational effectiveness may be further increased through augmentation by a civilian law-enforcement officer who would be responsible for search, seizure, and arrest actions. Active-component units are permitted to assist in marijuana eradication in support of domestic law-enforcement agencies (DLEAs). Marijuana-eradication operations are an application of the MP AS, mobility, and L&O functions, thus providing realistic training while supporting an operation.

11-21. As discussed in previous chapters, the MP are well suited to augment customs operations. They routinely perform customs inspections for US forces in Korea and Germany. The NG MP regularly conduct vehicle, cargo, and container searches under the supervision of US Customs agents and at US ports of entry.

11-22. The USACIDC agents can provide support to counterdrug operations. The CID units identify the sources of illegal-drug distribution systems. When directed, the MP and the CID complement interagency counterdrug efforts to stop the production, flow, and distribution of drugs and to provide planning, training, equipment, and facilities to support domestic counterdrug operations. The CID also supports CRIMINTEL programs, internal security, and site surveys at borders where Army units will deploy. In addition to the aforementioned support, the DOD General Counsel has ruled that MWD teams can be loaned to DLEAs to assist in detecting smuggled contraband.

SUPPORT IN COMBATING TERRORISM

11-23. The Department of Justice (DOJ) is the lead federal agency for dealing with acts of terrorism within the US, its territories, and its possessions. Within the DOJ, the Federal Bureau of Investigation (FBI) has the lead. The FBI can train the police of friendly nations in antiterrorism and counterterrorism operations. Usually, US military forces act in a technical-advisory but not a law-enforcement capacity in combating acts of terrorism. It is often difficult to distinguish the acts of politically motivated terrorists from violent acts of criminals or individuals in society at large. In fact, all terrorist acts are criminal; however, they also may be part of an insurgency.

11-24. The MP are key players in the defensive planning, implementation, and control of combating-terrorism operations, whether in garrison or deployed to war or MOOTW. In whatever countries US forces operate, MP units conduct antiterrorism operations daily as part of their L&O and AS functions. The MP also respond to terrorist incidents. They help secure the area and search for evidence, and they help the lead agency. The MP provide advice and assistance in developing antiterrorism and counterterrorism programs for deployed US forces and DOS agencies and for police, paramilitary, and military forces of assisted nations (when so authorized). MP training, liaison, and joint patrols help strengthen HN-police organizations, improving their effectiveness and efficiency in combating terrorism. The MP also assist HN police in performing PRC operations.

11-25. The training of police (domestic and foreign) requires DA approval, and restrictions exist on some forms of training. Legal advice and coordination are a must before training can occur. The MP and the CID provide personnel and equipment for planning, advising, equipping, and training agencies to prevent, deter, and respond to terrorism. The training focuses on deterring terrorist acts such as bombings, kidnappings, assassinations, taking hostages, and hijackings. Related assistance may include training and equipment for explosive detection, management of hostage situations, physical security, protection of key personnel, and defensive/evasive driving methods. For counterterrorism operations on US military installations, the MP provide highly trained, specially equipped SRTs for situations requiring the selective use of force. Additionally, CID agents are highly trained in hostage negotiation, which increases the commander's options before resorting to force.

11-26. The MP and the CID assist in counterterrorism information-gathering activities through liaison and joint patrols with HN police. These activities are part of the PIO function. The MP have extensive contacts with civilian police, who in turn have contacts with the populace. These contacts can help surface criminal information concerning terrorist plans, recruitment, and support structures sooner than military sources. The CID can provide commanders with CRIMINTEL on a daily basis. At the same time, MP antiterrorism support includes measures to reduce the vulnerability of people and property. Such measures include—

- Personal awareness.
- Personal-protection operations and techniques.
- Crime prevention.
- Physical-security programs.

11-27. The MP and CID crime-prevention programs employ proactive measures to protect people and their property and US and HN property. Crime prevention is a continuous process of planning, implementing measures, evaluating effectiveness, modifying measures, and developing further information. The primary crime-prevention tool is the crime-prevention survey. It is used to examine the physical features of various critical sites. The MP physical-security programs help identify, reduce, eliminate, or neutralize conditions favorable to criminal, terrorist, and insurgent activities. The key measures used to accomplish this are physical-security inspections and surveys.

SUPPORT TO NONCOMBATANT EVACUATION OPERATIONS

11-28. Noncombatant evacuation operations (NEOs) relocate civilians who are endangered in a foreign nation. The NEOs are classified as benign (unopposed) or hostile (opposed). MP roles in support to NEOs differ accordingly. In a benign environment, NEOs are usually conducted with the help and full cooperation of the affected nation, with little or no opposition to evacuation. The use of force is limited to self-defense (as outlined by the ROE) and protection of the evacuees. In a hostile environment, NEOs are opposed by forces with specific intentions of preventing or destroying the operation. In this scenario, the MP will follow the ROE as established by higher headquarters.

11-29. MP operations in NEOs closely parallel EPW/CI evacuation operations. The objective of both is to safely collect, secure, protect, transport, process, and account for people, whether they are EPWs or noncombatants. However, due to the dissimilarities of these groups, the techniques employed differ.

11-30. The MP can secure selected facilities, reception and transfer points, or other locations required to collect, process, and evacuate people. They can provide route reconnaissance and mobile security teams to escort HRP, DOS personnel, evacuees, and sensitive material. If needed, MP units assist civil-affairs personnel with the movement and control of dislocated civilians. The MP may support civil affairs by setting up and operating evacuation sites, holding areas for foreign nationals denied evacuation, and reception and processing stations. This support can be provided by MP CS units for a short duration and by MP I/R units for a prolonged duration. Additionally, the MP will establish and maintain liaison with HN police to ensure a mutual understanding of jurisdiction, to coordinate efforts, and to provide criminal and tactical intelligence. In hostile environments requiring the employment of combat forces, the MP would be employed in their traditional CS role.

SUPPORT TO SECURITY-ASSISTANCE OPERATIONS

11-31. Security-assistance operations are one of the main tools of US foreign policy. A critical concern in these operations is the HN's ability to plan and manage its own defense resources. In security-assistance operations, the US provides defense materiel, military training, and defense-related services through grants, loans, credit, or cash sales. When authorized, the MP support security assistance with military and police training through the International Military Education and Training Program (IMETP). This program provides instruction and training to foreign military and civilian personnel on a grant/aid basis. Many foreign civil law-enforcement agencies request training in the following areas:

- Basic police procedures (such as patrolling and crime-scene protection).
- Physical security.
- Corrections.
- Civil-disturbance operations.
- Customs operations.
- Traffic control.
- Use of force.

11-32. The MP possess the expertise and experience to advise, train, and assist a HN's military and police forces. The MP's participation in the IMETP includes—

- Formal and informal instruction of foreign students.
- Technical education and tactical training with applicable training aids.
- Assistance to foreign police and military elements by MP mobile training teams (MTTs).
- Providing MP units or small, modular, functionally specific teams to support security-assistance needs. These teams can train HN police in both field and institutional facilities.
- Support to nation-assistance operations.

11-33. Nation-assistance operations support a nation's effort to promote development, ideally by using its national resources. This is achieved through education and the transfer of essential skills to the HN. The goals of such operations are to—

- Promote long-term regional stability.
- Develop sound and responsive democratic institutions.
- Develop supportive infrastructures.
- Promote an environment that allows for orderly political change and economic progress.

11-34. The MP support nation-assistance operations with training and joint MP and HN-police operations. The MP know and understand police operations, military operations, and C^2 procedures. Consequently, the MP are ideal for training the law-enforcement personnel in both field and institutional settings. The MP's nation-assistance goals are to—

- Enhance L&O in democratic societies.
- Improve efficiency and effectiveness.
- Promote the proper usage and maintenance of equipment.
- Establish a sound training base for police.

- Standardize procedures that enhance combined police operations.
- Promote friendship and goodwill toward the US.
- Preclude the need for US military interventions to counter acts of violence.

11-35. Combined police operations reinforce training and provide HN police with mobility, security, and communications to operate more effectively while conserving critical personnel resources. The MP provide support through training and the following:

- Law-enforcement, security, and criminal-information support to the HN police force (to include L&O operations and administration in a democratic society).
- Patrol and desk operations.
- Circulation-control operations.
- PRC and civil-disturbance operations.
- MWDs.
- Physical-security operations.
- Personnel-security operations.
- Area- and route-security operations.
- Counterdrug operations.
- Antiterrorism operations.
- Mass immigration operations.
- Customs and border operations.
- Confinement operations.
- Crime prevention.
- Community and police relations.
- C^2 of police operations.

11-36. When conducting nation-assistance operations, as well as other operations, the MP will often be tasked to support civil-affairs operations. Close coordination with the G5/S5 and liaison with the appropriate HN representatives have added emphasis to these types of operations. The scope of civil-affairs operations is influenced by the economic, social, psychological, and political background of the country and its people. Civil-affairs personnel will coordinate the MP's effort to restore stability, contribute to national development, and promote support for the host government. Civil-affairs personnel assist in preventing civilian interference with military operations and coordinate all other military and civil affairs (such as community relations, PRC, civil defense, and community- and area-security programs).

11-37. The MP can conduct ADC measures to assist the HN in the aftermath of a natural or man-made disaster. They help rescue and evacuate the injured for medical care, and they issue food, water, and essential supplies (See FM 41-10).

SUPPORT TO PEACEKEEPING OPERATIONS

11-38. Peacekeeping operations (PKOs) are military operations conducted with the consent of the belligerents in a conflict. The PKOs are intended to maintain negotiated truces and to facilitate diplomatic solutions. The US may participate in a PKO under the auspices of an international organization, in cooperation with other countries, or unilaterally. A PKO usually occurs in an ambiguous situation in which there is extreme tension and the possibility of violence. This violence may range from terrorism and sabotage to minor military conflicts involving known and unknown belligerents. MP forces may serve as an element of a UN peacekeeping force or as part of a multinational observer group. The overall operational control of a peacekeeping force is exercised by a multinationally staffed military peacekeeping command. However, the commander of each assigned national force retains command integrity of his unit.

11-39. The basic MP force structure and appropriate augmentation are situation dependent. The size and composition of the MP forces are determined by diplomatic negotiations. Personnel spaces are allocated based on the US contingents' respective missions. The principle of consent affects the composition and function of the peacekeeping force. Consent applies to the degree of interest the disputing countries have when contributing their forces to participate in the peacekeeping effort.

11-40. Each PKO is unique but is generally characterized by constraints in the use of force. In a PKO, force is normally limited to self-defense, the defense of others, and the protection of national security. Therefore, the soldiers in a PKO must possess the skills required for warfare, but they must also be trained so that they are equally able to act as intermediaries. In this aspect, the MP are a distinct asset. They are experienced in a case-by-case approach to the use of force. They are experienced in using discretion and in the de-escalation of potential violence. The MP training in EPW operations are especially useful in handling EPWs if the peacekeeping force supervises the exchange of prisoners of war.

11-41. The peacekeeping force employs a combination of the following techniques to control the potential for violence inherent to PKOs:

- Observation is the peacekeeping force's primary responsibility and basic requirement. It provides timely and accurate reports of any suspicious situation, incident, or occurrence.
- Information gathering is a continuous requirement. Peacekeepers must be constantly alert to what takes place around them and to any changes or inconsistencies in the behavior, attitude, and activities of military and civilian personnel.
- Surveillance and supervision operations help oversee the implementation of agreements. The operations require restraint, tact, and patience.
- Complaint investigations must be thorough and objective, and they must result in a fair assessment.
- Negotiation and mediation skills are often required of peacekeepers. If peacekeepers can resolve minor problems at the lowest level, they can prevent major issues from rising, which may prevent increased tensions or the resumption of fighting.
- Patrolling (mounted and dismounted) among the population publicizes presence. It also serves as a means of gathering information, informally enhancing supervision, and investigating complaints.

11-42. The above techniques are compatible with MP battlefield functions. Operationally, the MP conduct route and area R&S, operate mounted or dismounted patrols, operate listening/observation posts, provide humanitarian assistance, investigate possible terrorist and criminal acts and cease-fire and sanction violations, and gather information. The MP provide a flexible, wide-ranging means of information dissemination. Additionally, they provide crucial support for force protection. The MP and the CID employ active and passive measures to deter and defeat threats directed at service members, their families, DOD civilians, facilities, equipment, and very important persons (VIPs). The MP use emerging technologies in force protection to conduct security, detection, and identification operations and to warn of intruders.

11-43. As part of a PKO, the MP conduct crowd- and riot-control operations, including the extraction of mob leaders. The MP are an ideal force for controlling antagonistic masses engaged in rioting, looting, and demonstrating. The MP understand how to make the transition from the lower end of the use-of-force spectrum to the use of lethal force, if so required or directed. Finally, the MP deploying with peacekeeping forces can provide early-on force protection, including headquarters and initial aerial port/ seaport of debarkation (APOD/SPOD) security. The MP's mobility, firepower, and communications provide critical reconnaissance, information-collection, and response-force capabilities and acceptability not found in combat forces.

11-44. When conducting crowd- and riot-control operations in a PKO or peace-enforcement operation (PEO), nonlethal weapons (NLWs) are an additional means of employing force for the purpose of limiting the probability of death or serious injury to noncombatants or belligerents. However, the use of lethal force must always remain an inherent right of individuals in instances when they, their fellow soldiers, or personnel in their charge are threatened with death or serious bodily harm. The NLWs add flexibility to the control of disturbances by providing an environment where military forces can permissively engage threatening targets with limited risk of noncombatant casualties and collateral damage (see FM 90-40).

11-45. The use of lethal force (employed under the standing ROE) will never be denied. At no time will forces be deployed without the ability to defend themselves against a lethal threat nor will they forego normal training, arming, and equipping for combat. Nonlethal options are a complement to, not a replacement for lethal force. They seek to expand a proactive response across the range of military operations.

11-46. The decision to use NLWs against an adversary during a confrontation should be delegated to the lowest possible level, preferably to the platoon or squad. However, this requires that all personnel (not just leaders) have a clear understanding of the ROE and the commander's intent.

11-47. Commanders and public-affairs officers must be prepared to address media questions and concerns regarding the use and role of NLWs. They must be prepared to address the role of NLWs, and they must make it clear that the presence of NLWs in no way indicates abandoning the option to employ deadly force in appropriate circumstances.

11-48. The PEOs are in the high end of the PKO's spectrum. These operations are just short of combat and require a different operational approach. MP operations to support PEOs correspond to MP battlefield functions supporting combat forces. The PEOs often pose special operational considerations for dealing with dislocated civilians. The MP provide PRC operations and I/R, when required.

ORGANIZATIONS AND CAPABILITIES

11-49. MP units are well suited for stability and support operations given their unique capabilities and low force signature. The MP brigade, battalion, company, and I/R units may be tailored for stability and support operations based on their unique METT-TC.

MP BRIGADE TF

11-50. In essence, the organization and capabilities of the MP organization in stability and support operations is the same as those for other operations. However, its augmentation is tailored based on METT-TC unique to stability and support operations and the commander's needs. The MP brigade has been the centerpiece of successful MP deployment during previous stability and support operations. The MP brigade provides comprehensive planning, C^2, and A/L support to assigned and attached subordinate units. The brigade also has the command and staff experience to coordinate and supervise HN security and law-enforcement activities.

11-51. The MP brigade headquarters provides senior MP leadership and the communications capabilities required to direct a variety of functional elements—CSS, PSYOP, public affairs, and combat-arms augmentation—in support of joint or unilateral TF operations. At authorized level of organization (ALO) 1, the MP brigade headquarters has an SJA section to coordinate the legal and jurisdictional issues inherent to stability and support operations. The brigade headquarters also has an S5 section and a liaison capability to coordinate support and assistance to HN agencies, friendly forces, other US forces, DOS personnel, and nongovernmental or private volunteer organizations.

11-52. The MP brigade TF (MPBTF), with its robust capabilities and low force signature, is well suited for the unique mission requirements and sensitivities of stability and support operations. The MPBTF can be employed as part of a combat TF in high-threat, opposed-entry operations or as the lead TF for a low-threat, unopposed-entry operation. The following are examples of previous MPBTFs and the operations in which they participated:

- **Operation Provide Comfort.** The 18th MP Brigade deployed to northern Iraq to provide humanitarian assistance and protection to the Kurds.
- **JTF Los Angeles.** The 49th MP Brigade (NG) provided civil-disturbance assistance to Los Angeles in the aftermath of the Rodney King trial.
- **Operation Hawkeye.** The 16th MP Brigade and units of the 89th MP Brigade provided security, protection, and humanitarian

assistance to the population of Saint Croix in the aftermath of Hurricane Hugo.

- **JTF Andrew.** The 16th MP Brigade provided humanitarian assistance in the aftermath of Hurricane Andrew.
- **Operation Sea Signal.** The 89th MP Brigade deployed to Guantanamo Bay, Cuba, to provide humanitarian assistance to Haitian migrants.
- **Operation Restore Democracy.** The 16th MP Brigade deployed to Haiti to provide humanitarian assistance.

11-53. Early deployments of the MPBTF headquarters or a mission-tailored portion of the headquarters positions a senior, effective MP C^2 headquarters to respond to all MP requirements. It can assist in the coordination with the media, governmental and nongovernmental agencies, and other forces, and it can coordinate civil-military aspects of the operation. The MP brigade headquarters can assimilate and employ additional forces without degrading operations. The augmentation of brigade headquarters is driven by METT-TC. With the brigade headquarters in place, subordinate or supporting units can be significantly expanded without the loss of C^2. In essence, little or no loss in operational momentum occurs when the C^2 element deploys first.

11-54. The MP LRP section of the MP brigade headquarters colocates and works with the higher headquarters' G3/Operations Directorate (J3) plans element. This allows the MP LRP to incorporate MP operations effectively into higher-level planning, and it enables an efficient transition to field operations. Also, because in peacetime this section functions as a PM cell, it provides the PM with efficient and effective technical support during stability and support operations. As the operations approach termination or enter sustainability and fewer forces are required, the brigade headquarters can redeploy, leaving a battalion TF in charge of MP forces.

MP BATTALION TF

11-55. The MP battalion (CS) provides command, control, and support for assigned MP companies (CS). MP battalions (CS) are best employed as part of a TF—either as part of a combat TF (high threat, opposed) or a support TF (low threat, unopposed). With sufficient augmentation, the battalion can provide the logistical, administrative, personnel, and operational support needed to lead a diverse TF. During stability and support operations, an MP battalion TF may be placed under the OPCON of a brigade-size TF (MP or non-MP) or a division TF. If placed under the OPCON of a division TF, both the division PM and the MP battalion commander will synchronize their efforts to maximize the employment of available divisional and nondivisional MP assets in support of the division commander's concept of operations.

11-56. As with the MPBTF, the MP battalion's contributions to stability and support operations are not a new concept. The following are examples of previous battalion TFs and the operations in which they were involved:

- **Cuban Refugee Movement.** The 519th, 716th, 720th, and 759th MP Battalions deployed for humanitarian-assistance operations.
- **Operations Restore Hope and Continued Hope.** The 720th MP Battalion provided humanitarian assistance in Somalia.

- **Operation Sea Signal.** The 720th MP Battalion deployed to Guantanamo Bay, Cuba, to conduct humanitarian-assistance operations for Haitian migrants. Later, the 716th and 759th MP Battalions also deployed to Guantanamo Bay to perform like functions for the Cubans.

MP COMPANY IN STABILITY AND SUPPORT OPERATIONS

11-57. The MP company is the MP Corps's primary mission performer. It executes a wide variety of missions. In these operations, MP companies—

- Provide mobility support for both vehicles and personnel.
- Provide AS, including security for critical personnel, sites, cargoes, and railways.
- Secure, safeguard, shelter, and control detainees and dislocated civilians.
- Restore order in civil disturbances and conduct PRC operations.
- Carry out limited L&O operations.
- Provide liaison, coordination, and training for all aspects of law enforcement to other agencies.
- Conduct police-intelligence collections and disseminate information to the populace, friendly forces, and other HN agencies.
- Support force protection and security.

11-58. The MP company possesses robust mobility, lethality, and communications. Each of the MP companies' three-man teams are equipped with either a high-mobility, multipurpose, wheeled vehicle (HMMWV), an up-armored HMMWV, or an armored security vehicle (ASV); an AN/VRC 91 radio; a Light-Vehicle Obscuration Smoke System (LVOSS); an MK-19 grenade machine gun; a squad automatic weapon (SAW); two M-16 or M-4 rifles; an M203 grenade launcher; and a 9-millimeter pistol as a side arm for each team member. In addition to conducting combat operations, the MP company commander has the capability to conduct limited L&O operations (desk and investigative operations). With a tremendous operational reach, the MP company will displace its platoons as far as possible to accomplish all of the MP functions.

MP I/R UNITS IN STABILITY AND SUPPORT OPERATIONS

11-59. The MP I/R units are specifically equipped and trained to support, safeguard, account for, and provide proper and humane treatment for EPWs/CIs, US military prisoners, or dislocated civilians. When conducting dislocated-civilian operations, the MP I/R units support and interact with civil-affairs and PSYOP units who are responsible for monitoring and handling dislocated civilians. Should more than one category of personnel have to be supported, the MP I/R units will establish separate compounds for each to preclude contact with other populations. When operation of the facility is going to be more than a short-term endeavor, the intent is to involve (as appropriate) NGOs, IHOs, HN authorities, and dislocated civilians in the administration of the facility. The tasks performed by MP I/R units for stability and support operations are basically the same as those for war, but the operations and procedures are modified according to the operational environment and the categories of people to be housed in the facility (see FM 19-40).

11-60. Criminal detainees (who are detained against their will) are subject to controls and procedures that closely parallel those used for EPWs/CIs. The primary emphasis is on guarding and preventing escape until their disposition is determined by recognized civilian judicial/law-enforcement authorities. When the operation is conducted outside of the US and its territories, criminal detainees are treated according to established legal procedures.

11-61. Dislocated civilians (who are being assisted rather than detained) are provided aid, shelter, and protection. The emphasis is on protecting them from harm by natural forces or hostile personnel. A special category of personnel arises when the MP I/R unit is required to house personnel that are dislocated civilians but who must be detained against their will. Such is the case of mass migrants who flee their countries and find themselves under US custody while formal proceeding guidance is being developed. In this case, MP I/R units must be sensitive to the situation and attempt to strike a balance between security, shelter, protection, and detention procedures.

11-62. Military detainees (who are detained against their will) are subject to controls and procedures that closely parallel those used for EPWs/CIs. In stability and support operations, US forces detain opposing-force military personnel. Without a formal declaration of war, opposing military personnel who are captured cannot be categorized as EPWs. However, they are afforded many of the EPW's rights and privileges under the Geneva convention. Such a situation occurred during Operation Just Cause when PDF personnel were detained by US forces.

11-63. In an operational environment in which hostile groups are engaged against one another, an I/R facility may be set up to protect one group from another. In this case, the facility's purpose is to shelter, sustain, account for, and protect its dislocated civilians from the hostile group. The MP I/R units and other supporting units concentrate on providing AS to protect the facility from direct fire. Other MP or combat forces provide protections beyond the direct-fire zone. Accountability for the dislocated civilians is coordinated with the SJA and civil affairs. They focus on maintaining a record of the people in the facility and their physical conditions. In operations in which no hostile groups are engaged (such as natural disasters), the I/R facility may be set up to provide shelter, sustain, and account for personnel. There may not be a need for external security personnel.

11-64. The MP I/R C^2 structure for stability and support operations is based on METT-TC. The nature and complexity of the mission, the number and types of detainees, and the operational duration should be considered. At one end of the operations spectrum, an MP brigade (I/R) may be required; while at the other end, an MP battalion (I/R) may be assigned to an MP brigade (CS).

Chapter 12

Force Protection

There is no more important responsibility than force protection. It is an inherent command responsibility and must be fully integrated into every unit's mission. A commander must continually review his unit's force-protection posture and crosswalk it with current and changing policy and threat levels. Force protection demands the personal involvement of commanders to ensure the best possible security consistent with the threat to personnel and mission-essential resources.

AR 525-13

Force protection consists of those actions that prevent or mitigate hostile actions against DOD personnel (to include family members), resources, facilities, and critical information. It coordinates and synchronizes offensive and defensive measures to enable the joint force to perform while degrading opportunities for the enemy. It includes air, space, and missile defense; NBC defense; antiterrorism; defensive-information operations; and security to operational forces and means.

OVERVIEW

12-1. The MP contribute significantly to the Army's overall force-protection effort. They are directly responsible for executing certain security-related measures (law enforcement and physical security) and for providing support to others (HRP security and antiterrorism), all of which contribute to force protection. Commanders use the Army's risk-management methodology and established security programs and capabilities within US forces to help prioritize and implement force-protection measures.

SUPPORT TO FORCE PROTECTION

12-2. MP support to force protection is accomplished under the leadership of the respective command's PM. It is accomplished mainly through an active role in physical security, HRP security, law enforcement, and antiterrorism.

PHYSICAL SECURITY

12-3. Physical security is an integral part of security of forces and means. Physical-security measures deter, detect, and defend against threats from terrorists, criminals, and unconventional forces. These measures include—

- Fencing and perimeter standoff space.
- Lighting and sensors.

- Vehicle barriers.
- Blast protection.
- Intrusion-detection systems and electronic surveillance.
- Procedural measures such as security checks, training and awareness, property accountability/inventory requirements, physical-security inspections of mission-essential or vulnerable areas (MEVAs), and physical-security surveys of installations.

12-4. These physical-security measures are not exclusively peacetime measures. They are designed to protect US personnel and equipment regardless of the mission or the geographical location.

12-5. The MP support the commander's security of forces and means programs by—

- Controlling or monitoring installation, base, or base-cluster access or entrance points.
- Monitoring intrusion-detection systems and providing a response force.
- Conducting physical-security inspections.
- Conducting perimeter security or site surveillance.
- Recommending the placement of walls, berms, gates, or barriers around designated MEVAs, high-value areas, or perimeters.
- Supporting the commander's risk-analysis effort.
- Conducting roving patrols, checkpoints, or roadblocks.
- Performing other physical-security measures as required by the commander.

HRP SECURITY

12-6. The HRP security program provides additional security to designated individuals (and their families) whom by virtue of their rank, assignment, symbolic value, vulnerabilities, or location or a specific threat are at a greater risk than the general population. This security program is applicable across the full spectrum of military operations.

12-7. The MP support the commander's HRP security measures by—

- Conducting protective-service details for HRP (normally conducted by USACIDC special agents).
- Responding to duress alarms and intrusion-detection systems.
- Conducting a PSVA (performed by the USACIDC).
- Employing MWDs.
- Coordinating with local, state, and federal agencies as well as joint, multinational, and HN law-enforcement agencies.
- Performing other HRP security measures as required by the commander.

LAW ENFORCEMENT

12-8. The Army's law-enforcement program supports the security of forces and means through the prevention, detection, response, and investigation of crimes. The MP, the USACIDC, and DOD police play a crucial role in the law-enforcement program. They provide a visible deterrent against a broad portion of the threat spectrum, investigate crimes, and provide an initial response force to threat incidents. As with the previous programs, the law-enforcement program is applicable anywhere across the spectrum of military operations.

12-9. The MP support law-enforcement operations by—

- Providing liaison teams with local, state, and federal agencies; HN police; and joint and multinational agencies.
- Employing SRTs and hostage-negotiation teams (normally provided by the USACIDC).
- Providing traffic enforcement, MSR regulation enforcement, and other route-control measures.
- Employing MWDs.
- Conducting MP investigations (criminal and noncriminal).
- Conducting patrolling, AS, and surveillance measures.
- Implementing applicable threat-condition (THREATCON) measures.
- Conducting and implementing other law-enforcement measures as required by the commander.

ANTITERRORISM

12-10. Terrorism is characterized as the unlawful use of violence or the threat of violence to coerce or intimidate a government or a society. Protection against a terrorist threat requires both an offensive counterterrorism capability and a defensive antiterrorism program. The primary focus of antiterrorism is on training and awareness and thorough planning. Antiterrorism is not a stand-alone program. It leverages other force-protection-related programs (such as physical security and law enforcement) to provide much of the physical defense against terrorism.

12-11. The MP support the commander's antiterrorist program by—

- Collecting CRIMINTEL (within the limits of AR 190-45).
- Responding to terrorist threats and incidents.
- Performing aggressive patrols and R&S operations.
- Detecting suspicious activities.
- Enforcing joint and cooperative agreements with the FBI, the USACIDC, MI, and other pertinent agencies (home or abroad).
- Conducting information collection and dissemination.
- Protecting critical assets, facilities, and personnel.
- Performing other antiterrorism measures as required by the commander.

Appendix A

Metric Conversion Chart

This appendix complies with current Army directives which state that the metric system will be incorporated into all new publications. Table A-1 is a conversion chart.

Table A-1. Metric Conversion Chart

Metric to English			English to Metric		
Multiply	**By**	**To Obtain**	**Multiply**	**By**	**To Obtain**
Length					
Centimeters	0.0394	Inches	Inches	2.54	Centimeters
Meters	3.28	Feet	Feet	0.0305	Meters
Meters	1.094	Yards	Yards	0.9144	Meters
Kilometers	0.621	Miles (stat)	Miles (stat)	1.5609	Kilometers
Kilometers	0.540	Miles (naut)	Miles (naut)	1.853	Kilometers
Millimeters	0.039	Inches	Inches	25.40	Millimeters
Area					
Square centimeters	0.1550	Square inches	Square inches	6.45	Square centimeters
Square meters	10.76	Square feet	Square feet	0.0929	Square meters
Square meters	1.196	Square yards	Square yards	0.836	Square meters
Volume					
Cubic centimeters	0.610	Cubic inches	Cubic inches	16.39	Cubic centimeters
Cubic meters	35.3	Cubic feet	Cubic feet	0.0283	Cubic meters
Cubic meters	1.308	Cubic yards	Cubic yards	0.765	Cubic meters
Milliliters	0.0338	US liq ounces	US liq ounces	29.6	Milliliters
Liters	1.057	US liq quarts	US liq quarts	0.946	Liters
Liters	0.264	US liq gallons	US liq gallons	3.79	Liters
Weight					
Grams	0.0353	Ounces	Ounces	28.4	Grams
Kilograms	2.20	Pounds	Pounds	0.454	Kilograms
Metric tons	1.102	Short tons	Short tons	0.907	Metric tons
Metric tons	0.984	Long tons	Long tons	1.016	Metric tons

Appendix B

Doctrine and Training Literature

Users of this manual must understand the elements of doctrinal literature and their relationship to each other. The commonly used terms doctrine and tactics, techniques, and procedures (TTP) are interrelated and mutually supportive; yet, each term has its own usage, level of detail, and place in the hierarchy of doctrinal publications. Because there is a close interrelationship, publications overlap in what they provide to the user. Figure B-1 illustrates the relationship of these doctrinal terms and publications.

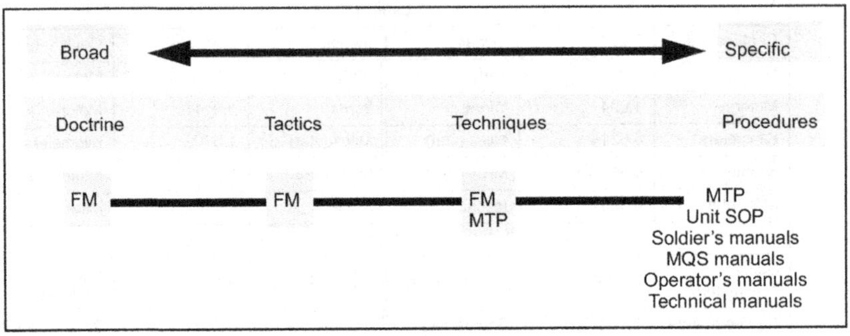

Figure B-1. Doctrinal and Training Literature

DOCTRINE

B-1. Doctrine is the body of fundamental principles by which the military guide their actions in support of national objectives. It is authoritative, but requires judgment in application. Doctrine provides the Army's fighting philosophy, establishes a common language, and structures the manner within which units plan and conduct combat operations.

B-2. Doctrine is not absolute; it is applied by the leader to meet the circumstances of the situation being faced. Doctrine provides for this flexibility with broad fundamentals and principles for conducting operations. Leaders well grounded in doctrine understand how to synchronize the elements of combat power and how to apply TTP using the assets available to accomplish the mission.

B-3. Doctrine is described in FMs. FM 3-0 defines the systems of war fighting practiced by the US Army. FMs 100-15 and 71-100 expand on this foundation. This manual describes the CS doctrine of the MP; it is the capstone manual for MP operations. It defines the role, operational requirements, functions, and C^2 relationships of MP units. This manual is not oriented on a specific organizational structure. Leaders at all levels apply this doctrine to their unique organizational and operational situations and use various tactics and techniques to accomplish the doctrine.

TACTICS

B-4. Tactics describe how the leader carries out doctrine (relating to the arrangement of forces for battle). Tactics include the—

- Employment of units in combat (to accomplish a mission assigned by the commander).
- Ordered arrangement and maneuver of units in relation to each other and to the enemy in order to use their full potential (a description of how the commander should arrange his forces and maneuver to accomplish a mission).

B-5. This is the part of doctrinal literature used for training in preparation for combat. Like doctrine, tactics presented in manuals are applied with judgment by the leader. The FMs prescribe "how to fight" and the mission training plans (MTPs) describe "what to train."

TECHNIQUES

B-6. Techniques are the methods of performing any act, especially the detailed methods used by troops or commanders in performing assigned tasks. Techniques describe the basic methods of using equipment and personnel. Techniques give details on how commanders actually carry out assignments. They improve the force's efficiency by ensuring the uniformity of action or by ensuring that the actions of various individuals and elements complement those of other individuals or elements.

B-7. The FMs and MTPs provide a description of techniques using the personnel and equipment available in actual organizations. These techniques show (at a more detailed level) how to use available assets to carry out tactics.

PROCEDURES

B-8. Procedures are the lowest level of detail. They address "how to" at the task level. Procedures are a particular course or mode of action that describe how to perform a certain task. Procedures include the—

- Standing methods used by units to accomplish tasks.
- Operating procedures for weapons and equipment.
- Crew drills.
- Staff actions and coordination requirements.
- Methods of target engagement by direct- and indirect-fire weapons.

B-9. Procedures are building blocks of individual- and collective-task accomplishment that serve as the foundation of tactics and techniques. Procedures are explained in unit standing operating procedures (SOPs), MTPs, soldier's manuals, operator's manuals, technical manuals (TMs), and similar publications.

Appendix C

Battlefield Workload Analysis

With minor adjustments in the capability block, the BWA can be used to determine MP requirements for the division. A BWA work sheet is shown in Figure C-1. A separate matrix is prepared for each EAC and corps MP brigade. Figure C-2, page C-3, is a sample of a BWA matrix for a 3-division corps MP CS company.

Mission/Function[1]	Capability[1]	Workload[2]	Rqmt	HN[18]	US Rqmt	Risk[19]	Mission Rqmt
Divisions supported[3]	1						
MMS	360 km MSR and ASR						
Refugees on control routes[4]	150,000 per day						
AS—corps rural terrain[5]	2,000 sq km						
AS—EAC rural terrain[5]	3,000 sq km						
AS—urban terrain[5]	800 sq km (urban)						
AS—USAF main opns base[5]	1,000 sq km/USAF MOB						
AS—convoy escort[6]	# tms per convoy	# convoys					
AS—ammo security[7]	1.3 modular ammo co						
	2 modular ammo co						
AS—pipeline security[8]	400 km						
AS—EAC pipeline co[8]	3						
AS—trains[9]	7						
AS—small critical sites[10]	12						
AS—major HQ security[11]	1						
AS—EAC port security[12]	0.66 of port						
AS—combat operations	When required, assets are diverted from other missions.						
L&O—incident response[13]	75,000 nondivision troops						
I/R—US prisoner detention[14]	700						
I/R—EPW escort/corps[15]	1,900 walking						
	2,500 by vehicle	Min: 1 plt					
	3,800 by train						
I/R—EPW holding/corps[16]	2,000	Min: 1 plt					
Other assigned missions[17]	Mission	Force size					
		Totals[20]					
Battalion HHD requirement = 0.199 x number of companies[21]							

Figure C-1. Sample BWA Work Sheet

NOTE: Workload ÷ capability = doctrinal requirement - HN offset = US requirement - acceptable risk = mission requirement

[1]Mission/capabilities are as stated in section I of TOEs 19477L000 and 19677L000. The unit capability for each mission is what the unit can accomplish when performing no other mission (for example, if it is only doing MSRs, the unit can cover 360 kilometers in a 24-hour period). Unless otherwise specified, the capabilities are the same at corps and EAC.

[2]Workload is the amount (MSR or area) or number of supported units/facilities.

[3]Corps support to each division is covered by allocating one MP company per supported division. Any additional support requirements should be based on a detailed division PM/MP mission analysis and included under other assigned missions.

[4]Unless specifically tasked, the Army normally relies on HN support to manage refugees. Redirecting refugees from MSRs and keeping them off of MSRs is part of (as is straggler control and TCP) the mission and MSR workload. When assigned the mission, the capability is for control of refugees on established refugee control routes that are in addition to the MSRs.

[5]Compute each square kilometer mission requirement separately. Do not compute the same terrain in more than one computation (for example, the area around the Air Force's main operating base (MOB) is not included with either rural or urban square kilometers).

[6]Convoy-escort capability is not stated in the TOE and is an AS function rather than an MMS function. Determine the average number of teams per mission and the average number of simultaneous missions. Multiply the number of teams required per mission by 0.028, then multiple the result by the number of simultaneous missions to determine the total requirement. State the number of teams per convoy and the number of convoys, then convert the force size to a 3-decimal-place company equivalent. Each additional team, squad, and platoon equivalent equals 0.028, 0.083, and 0.25 respectively.

[7]Security for modular ammo companies (ASPs) supporting divisions is three platoons and is normally conducted directly behind the division (corps forward). However, METT-TC may require the ammo units to move into the DRA. In this case, corps units should be attached to the division for the duration. Security for corps storage areas (CSAs) and theater storage areas (TSAs) is one MP company per two ammo companies. Security is provided on an area-support basis in the form of screening, with limited fixed posts.

[8]The EAC petroleum terminal/pipeline operating company manages the EAC pipeline that is normally laid adjacent to MSRs. Pipeline security is inherently provided by the MP operating on the MSRs. Unless the pipeline is not near the MSR, MP support is only allocated against the EAC petroleum terminal/pipeline operating company's bulk storage facilities. Security is provided on an area-support basis in the form of screening, with limited fixed posts.

[9]Train security is normally an EAC mission. This capability equates to simultaneously providing five MP teams to each of seven trains. Actual workloads and requirements are computed in the same manner as a convoy escort.

[10]Small critical facilities are normally defined as key MSR bridges and tunnels and communication nodes that may require additional security.

[11]The major headquarters are corps, COSCOM, TSC, unified command and, if directed, the JTF. The ASCC depends on the MP to provide security for two CPs. The ASCC requirement is an additional three platoons to support a second CP.

[12]The requirement for one seaport is 1.5 companies.

[13]Teams provide incident response, crime-scene security, and traffic control. They also assist L&O detachment teams and CID investigators.

[14]For deliberate planning in support of an OPLAN, personnel to be confined will normally be held by either the unit or the EAC I/R MP battalion performing the confinement mission. This does not preclude a requirement for the echelon PM to operate a field detention facility until prisoners are released back to their units or evacuated to the EAC confinement facility.

[15]The primary mode is vehicle (back-haul doctrine). Regardless of the mode, a minimum of one platoon (0.25 requirement) per supported division is allocated for EPW escort from the division central collection point (DCCP) to the corps holding area (CHA). Because the actual number of EPWs is unknown, coupled with no time/distance study for each supported division, one platoon per division provides a minimum capability for continuous evacuation within the stated doctrinal timeline (evacuate to the CHA within 12 hours of arrival at the DCCP).

[16]At a minimum, one platoon (0.25 requirement) is allocated to operate the CHA (provides the capability to hold 500 EPWs at any given time).

[17]Add any additional continuous missions assigned by the echelon commander not covered above. Examples include VIP escort/security for other than the CG and the deputy commanding general (DCG) of major headquarters, screening exposed corps flanks, and increasing teams per mission/function which decreases doctrinal capability (for example, permanent TCPs or checkpoints). Each additional team, squad, or platoon equivalent equals 0.028, 0.083, and 0.25 respectively.

Figure C-1. Sample BWA Work Sheet (continued)

[18]For each mission/function, any workload either fully or partially accomplished by the HN is subtracted from the doctrinal mission requirement. Enter the MP company equivalent offset and subtract from the doctrinal requirements to determine the US requirement.

[19]For each mission/function, if the operational/MP commander will accept mission risk, subtract the MP company equivalent to determine the mission requirement.

[20]Add columns to determine the total doctrinal company requirement, HN offsets, US requirement, acceptable risk, and mission requirement to support the operation/AO. Round the requirement to the nearest whole number (0.4 or fewer rounds down, 0.5 or greater round up).

[21]Determine battalion HHD requirements by multiplying the company requirements by 0.199 and rounding.

Figure C-1. Sample BWA Work Sheet (continued)

Mission/Function	Capability	Workload	Rqmt	HN	US Rqmt	Risk	Mission Rqmt
Divisions supported	1	3	3.000		3.000		3.000
MMS	360 km MSR and ASR	1,050	2.917		2.917		2.917
Refugees on control routes	150,000 per day	25,000	0.167	0.167	0.000		0.000
AS—corps rural terrain	2,000 sq km	6,500	3.250		3.250	0.500	2.750
AS—EAC rural terrain	3,000 sq km						
AS—urban terrain	800 sq km (urban)	1,500	1.875		1.875	0.500	1.625
AS—USAF main opns base	1,000 sq km/USAF MOB	1,000	1.000		1.000		1.000
AS—convoy escort	5 tms per convoy	4	0.560		0.560		0.560
AS—ammo security	1.3 modular ammo co	3	2.308		2.308		2.308
	2 modular ammo co	1	0.500		0.500		0.500
AS—pipeline security	400 km						
AS—EAC pipeline co	3						
AS—trains	7						
AS—small critical sites	12	9	0.750		0.750	0.250	0.500
AS—major HQ security	1	2	2.000		2.000		2.000
AS—EAC port security	0.66 of port						
AS—combat operations	When required, assets are diverted from other missions.						
L&O—incident response	75,000 nondivision troops	45,000	0.600		0.600	0.300	0.300
I/R—US prisoner detention	700						
I/R—EPW escort/corps	1,900 walking						
	2,500 by vehicle	Min: 1 plt/div	0.750		0.750		0.750
	3,800 by train						
I/R—EPW holding/corps	2,000	Min: 1 plt/	0.250		0.250		0.250
Other assigned missions	PSD for Corps CofS	1 sqd	0.083		0.083		0.083
	Flank screen	1 plt	0.250		0.250		0.250
		Totals	20	0.177	20	1.3	19
Battalion HHD requirement = 0.199 x number of companies			4		4		4

NOTE: Workload ÷ capability = doctrinal requirement - HN offset = US requirement - acceptable risk = mission requirement

Figure C-2. Sample BWA Matrix

Appendix D

Automated Information Systems in Support of Battle Command

As the MP Corps continues to advance into the twenty-first century, it must continue to pursue and stay in tune with the Army's initiative to acquire advanced technologies and operational concepts. This will give MP soldiers on the battlefield an information advantage that is critical for successful performance of their functions. The following information opens a window into the near future and explains some of the automation initiatives that will enhance the Army's and the MP Corps's soldiers.

ARMY INFORMATION SYSTEMS

D-1. The Global Command and Control System (GCCS) is the midterm solution and the bridge to the concepts outlined in the concept of command, control, communications, computers, and intelligence for the warrior (C4IFTW). The GCCS's core consists of the basic functions required by the war fighter to plan, execute, and manage military operations. The Army's selected C^2 system is the Army Battle Command System (ABCS).

D-2. The ABCS integrates the C^2 systems found at each echelon—from the ground-force component commander at the theater or JTF level to the individual soldier or the weapons platform. Whether ARFOR are deployed for land combat or are conducting peace operations, providing humanitarian assistance, or giving aid to civil authorities, the ABCS supports the mission by integrating the battle-space automation systems and communications that functionally link strategic and tactical headquarters. The ABCS is interoperable with joint and multinational C^2 systems at upper echelons across the full range of C^2 functionality, and it is vertically and horizontally integrated at the tactical and operational levels. The ABCS has three major components—

- The Global Command and Control System—Army (GCCS-A).
- The Army Tactical Command and Control System (ATCCS).
- The FBCB2.

GLOBAL COMMAND AND CONTROL SYSTEM—ARMY

D-3. The GCCS-A is the Army's component of the joint GCCS. It was built from application programs developed by the Army Worldwide Military Command and Control System [WMCCS] Information System (AWIS), namely the—

- Strategic Theater Command and Control System (STCCS).

- Theater Automated Command and Control Information Management System (TACCIMS).
- EAC portion of the Combat Service Support Control System (CSSCS) (CSSCS/EAC).

D-4. The GCCS-A is required in the MP LRP, S1, S3/NBC, and S4 sections within the MP brigade at EAC; in the Assistant Chief of Staff, G1 (Personnel) (G1), the G3, and the Assistant Chief of Staff, G4 (Logistics) (G4) sections within the MP command (I/R); and in the S1, S3, and S4 sections within the MP brigade (I/R).

ARMY TACTICAL COMMAND AND CONTROL SYSTEM

D-5. The ATCCS embodies the commander and his staff and the doctrine, procedures, and tools used to command and control forces on the tactical battlefield. The ATCCS's functional subsystems provide tools to support the C^2 process across the spectrum of low- to high-intensity conflicts worldwide; for MOOTW; and for combat, CS, and CSS functional areas at corps-and-below units. The ATCCS provides a common set of C^2 tools and functions for commanders that extends the ATCCS to the battalion level. The component's ATCCS's are the—

- Maneuver Control System (MCS).
- All-Source Analysis System (ASAS).
- Advanced Field-Artillery Tactical Data System (AFATDS).
- Air- and Missile-Defense Planning and Control System (AMDPCS).
- CSSCS.

D-6. The FBCB2 provides situational awareness and C^2 to the lowest tactical echelons. It facilitates a seamless flow of battle-command information and situation-awareness information across the battle space at brigade and below on Army platforms.

FORCE XXI BATTLE COMMAND—BRIGADE AND BELOW

D-7. The FBCB2 system, a battle-command information system, provides the technology to complete the ABCS information flow process from brigade to platform and across all platforms within the brigade TF. The FBCB2 enables commanders to remotely operate and maintain ABCS database connectivity (regardless of the command vehicle) and to digitally control and monitor their subordinate units' statuses and positions. Interoperable FBCB2 systems will enhance the battle commander's ability to have a detailed picture of the battlefield (situational awareness), to maximize force survivability, and to facilitate battlefield synchronization to reduce or destroy the enemy's capability to fight. The FBCB2 relies on integrating, processing, and moving information in a digital form. To support projections of future conflicts with an accompanying increase in the tempo of battle, the FBCB2 must be able to obtain information contained in databases from higher, lower, and adjacent elements without formally requesting that information through command channels. The war fighter requires the ability to pull data from all battlefield-system databases to keep current with the rapid changes caused by the dynamics of battle. The FBCB2 is required for divisional MP company commanders, PMs, platoon leaders, teams, and squads.

MANEUVER CONTROL SYSTEM

D-8. The MCS is the ABCS's horizontal and vertical integrator. The MCS will enable commanders and their staffs to collect, coordinate, and act swiftly on near-real-time battlefield information. It will quickly, accurately, and automatically move tactical information on the battlefield. This allows a commander to readily access information and display current situation reports (SITREPs) that assess enemy strength and movement as well as the status of friendly forces. The MCS will then aid the battle staff in rapidly disseminating the commander's orders. It is intended to provide two major functional capabilities—maneuver functional-area control and force-level information. The MCS is the ATCCS's functional subsystem that—

- Supports the combined-arms force commander directly by providing automated support for planning, coordinating, controlling, and using maneuver functional-area assets and tasks.
- Coordinates and synchronizes supporting arms in conducting operational planning, field operations, and training.

D-9. The MCS is required for division PM operations sections, company CPs at echelons corps and below (ECB), MP battalion commanders, S3 sections at corps, MP brigade commanders, and corps MP LRP and S3 sections.

ALL-SOURCE ANALYSIS SYSTEM—REMOTE WORK STATION

D-10. The ASAS—remote workstation (RWS) subsystem will operate within the existing intelligence and electronic warfare (IEW) battlefield operating system (BOS) and ABCS architectures to provide automation support to G2/S2 intelligence staff elements and other designated intelligence organizations from the maneuver battalion through EAC. It will operate at the security level appropriate to its echelon, mission, and functions. It assists the G2/S2 in providing IEW support to the commander and other staff elements in planning, coordinating, and conducting current and future tactical operations. It interfaces with the sensitive compartmented information (SCI) analysis and control element (ACE) subsystem to provide operational guidance and to exchange IEW information and products. It also operates as the technical control portion of the IEW C^2 node of the ABCS to provide current IEW and enemy-situation information to the force-level information database for access and use by ABCS users. Two RWSs are also provided to the brigade analysis control team (ACT) of the DS MI company to process data input from DS collection systems, to filter intelligence data from broadcast systems, and to pass tactically relevant enemy-situation information or situation-development data to the brigade S2's RWS. The ASAS-RWS is required in the S2 section of the CID group, the S2/S3 section of the CID battalion, the G2 section of the MP command (I/R), the S2 section of the MP battalion (I/R), and the S2 section of the MP battalion and brigade at all echelons.

COMBAT SERVICE SUPPORT CONTROL SYSTEM

D-11. The CSSCS will function as an integral part of the ATTCS. It is designated as an information system with the capability to rapidly collect, analyze, and disseminate logistical, medical, and personnel information to accelerate and improve decision making by CSS and force commanders. It also enables the rapid exchange of C^2 information with other battlefield functional areas (BFAs) through automated interfaces. The system will process unclassified through secret information. The CSSCS will be employed in all divisions, corps, EAC, separate brigades, and armored cavalry regiments. The MP require CSSCS functionality in all S1 and S4 sections in battalion or higher organizations except where the GCCS-A is required as stated above.

TRANSPORTATION COORDINATOR AUTOMATED MOVEMENT SERVER

D-12. The Transportation Coordinator Automated Movement Server (TC-AIMS II) is used by transportation agents and deploying units of each service to automate the processes of planning, organizing, coordinating, and controlling deployment and redeployment operations worldwide during peace as well as contingencies. The TC AIMS II will link all DOD-component unit movements with the installation transportation officer/traffic-management officer (ITO/TMO). This will provide the capability to plan and execute deployments rapidly to meet CINC-required arrival and departure dates. The TC AIMS II will also provide the ITO's/TMO's day-to-day traffic management functions and the theater transportation managers with the capabilities to use theater common lift assets effectively. The TC AIMS II will also allow CSS organizations and supporting commands to sustain the force and perform effective movement control, coordination, distribution, and sustainment once in theater. This is also defined as transportation movement and support of DOD personnel and cargo during all phases of military operations in all environments (including RSOI and battlefield operations). The TC AIMS II's functionality is required at all division PM operations and S3 sections within MP battalions and brigades to facilitate MSR regulation enforcement. Additionally, MP companies, battalions, brigades, and commands require the TC AIMS II to plan, organize, coordinate, and control deployments and redeployments.

GLOBAL COMBAT SUPPORT SYSTEM—ARMY

D-13. The GCSS-A is an integrated logistical support system with accompanying hardware that will connect the service's supply, finance, transportation, maintenance, and personnel databases. Under this system, the Army will integrate 13 Legacy tactical logistics systems and will eventually interface with the service's wholesale logistics systems as well as the consolidated logistics systems of the other services and the DOD. The system is critical for modernizing automation, for reducing logistical footprints in theaters, and for expanding war fighters' capabilities. The system will also replace the Standard Army Management Information System (STAMIS). The GCSS-A is required at all MP division company CPs and MP battalion and brigade S4 sections. Additionally, it is required at all supply functions within MP detachments that support I/R missions.

MILITARY POLICE AUTOMATED SYSTEMS

D-14. In an effort to maintain an information advantage and to acquire advance technologies and operational concepts, the MP Corps has acquired the following systems:

- The Military Police Management Information System (MPMIS) and its subsystems
- The CID's automated systems and their subsystems.

MILITARY POLICE MANAGEMENT INFORMATION SYSTEM

D-15. The MPMIS is a DA-mandated reporting system that does not interface with the ABCS. The MPMIS provides automated tools for MP records management and reporting. It reduces the administrative burden on the PM's staff. It increases the awareness of problem areas by managing crime information. The MPMIS consists of the following elements:

- **Offense Reporting Systems Windows (ORS-2W).** The ORS-2W provides automation support to track and report complaints, criminal incidents, and traffic incidents. It incorporates the CRC, the federally mandated National Incident-Based Reporting System (NIBRS), the DOD-mandated Defense Incident-Based Reporting System (DIBRS), the military community (MILCOM), and the installation's data requirements. The ORS-2W's functionality is required at PM activities at all echelons of command within the Army.
- **Correctional Reporting System (CRS-3).** The CRS-3 provides automation support to track and report inmates and corrections staff at all Army-controlled correctional facilities.
- **Security Management System (SMS).** The SMS provides automated support to assist physical-security inspectors and force-protection officers in reporting inspections and surveys. It tracks intrusion-detection devices and MWD training and health requirements.
- **Vehicle Registration System (VRS).** The VRS provides installations with an automated means to track vehicle, weapon, and bicycle registrations.
- **Prisoner-of-War Information System (PWIS-3).** The PWIS-3 provides the automated means of tracking an EPW's status from capture through release. This system is required at all EPW management activities.

CID AUTOMATED SYSTEMS

D-16. The Headquarters (HQ) USACIDC mandates CID automated systems. Similar to the MPMIS, it does not interface with the ABCS. The system facilitates data-collection efforts within the command. The CID's automated systems consist of the—

- **Automated Criminal Investigative Reporting System (ACIRS).** The ACIRS supports management and data-collection functions at the CRC and HQ USACIDC. It supports the following:
  - Maintaining investigative data.
  - Conducting agent investigative activities.
  - Reporting investigative data.
  - Managing the investigative processes.
  - Conducting criminal-activity analyses.

- **CRIMINTEL.** Data sources for CRIMINTEL include MP reports, CRIMINTEL reports and bulletins, reports of investigation, crime-prevention surveys, case files, collection folders, raw-data folders, significant data files, and target-analysis files. The CRIMINTEL interfaces with the ACIRS, but it does not share CRIMINTEL products.

Appendix E

Police Information Assessment Process

The PIAP differs somewhat from the tactical IPB. While the IPB allows the maneuver commander to see the effects that a variety of factors may have on his forces, the PIAP looks beyond this approach to determine how other relevant information may impact his forces. This information may be obtained from police or criminal actions or incidents encountered during the performance of MP functions.

IMPLEMENTATION

E-1. The PIAP is not a substitute for the IPB. MP leaders must continue to rely on the IPB and to use its estimates as a starting point for the PIAP. However, the PIAP is a dynamic and continuous cycle that complements the IPB (Figure E-1). The following paragraphs represent the doctrinal approach to implementing the PIAP. These steps provide the basic foundation to the process. They are not exclusive; the successful development of the PIAP will depend on the MP leader's ability to apply the process to his specific environment, METT-TC, and the commander's priorities.

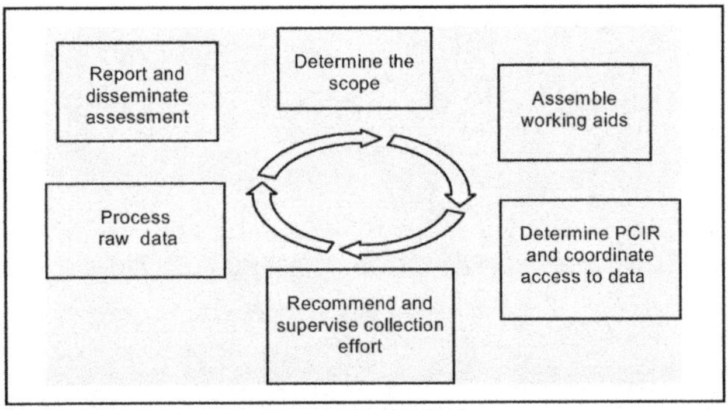

Figure E-1. The PIAP

DETERMINE THE SCOPE OF THE PIAP

E-2. While determining the scope of the PIAP might sound simple, it is one of the most crucial elements of the entire process. It entails identifying what you want to be the end result of your PIAP. To help determine the scope of the PIAP, MP leaders must perform the following subtasks:

- **Conduct a mission analysis.** A detailed mission analysis will ensure that all available resources are focused toward one goal. The more specific the mission is, the more focused and productive the effort. This involves more than just looking at the mission statement, the commander's intent, and the concept of operations. It involves translating the mission and situation into easily understood, manageable goals. The following are examples of these missions:
  - Identifying criminal threats from local national to US Army logistics operations within the port of Mogadishu, Somalia.
  - Determining the impact of gang violence on US forces operating in the vicinity of Camp Bondsteel, Kosovo.
  - Determining how rioters' violence will affect force protection before deploying forces to Los Angeles, California.
  - Determining the status of public service agencies (such as the police, the fire department, and the emergency medical services [EMS]) in Homestead, Florida, after the city was struck by Hurricane Andrew.

NOTE: As you can see from the examples, these missions are quite specific. In all likelihood, it would be necessary to establish several different PIAP missions to adequately protect Army elements operating in the MP's AOR.

- **Prioritize missions.** MP leaders must prioritize each PIAP mission since MP resources will probably be limited. The commander's intent is the key factor for determining priorities. When the commander's most important mission has been determined, MP leaders must decide how they can contribute to the overall success of that mission.
- **Determine the AO.** Generally, the AO is a geographical area (including the airspace above) usually defined by lateral, forward, and rear boundaries assigned to a commander. MP leaders must know where major units are located and how the characteristics of the location may impact ongoing operations. For example, units in direct contact with the enemy or isolated from the local population are less likely to be the target of a PIAP mission than those units adjacent to great population centers or in rest and relaxation centers.
- **Determine key terrain.** Key terrain is any locality or area that affords a marked advantage to the retaining side.
- **Determine the AOI.** An AOI is the area of concern to the commander. This includes the area of influence and the areas adjacent thereto and extending into enemy territory and to the objectives of current or planned operations. A thorough understanding of the AOI may help predict potential threat actions. For example, how far are local agitators and ringleaders willing to travel to interfere with US operations in Pristina, Kosovo? Would they go to other US compounds?
- **Determine required information products/sources.** The desired end-state product must be decided. Should it be a report? An updated SITREP? An upgrade in force-protection measures? Or a combination of products? It should also be determined who can provide the best

information (the CID; MI; the local police chief; or joint, interagency, or multinational agencies).

ASSEMBLE WORKING AIDS

E-3. During this step, MP leaders assemble the necessary graphic and nongraphic aids and information that will help them accomplish their mission. The CID and the MP may form an analysis team to track and analyze police information that might assist tactical operations or population-control operations. Working aids include, but are not limited to—

- **Maps.** Besides the standard military maps that depict the military AOs, the MP must attempt to procure indigenous maps of AOs. Detailed city and street maps will prove to be invaluable even if they are in a foreign language.
- **Language aids.** Phrase books and translation dictionaries cannot take the place of intensive language training, but they are better than nothing. These are essential even if the MP have assigned interpreters.
- **Open-source information.** This can cover an extensive range of invaluable information, but it should be available from a variety of sources. Examples include—
  - Demographic information on the HN population.
  - Societal information on the HN.
  - The locale and disposition of HN police forces.
  - Historical crime data of the area.
  - Environmental information (terrain and weather factors).
  - Local newspaper articles.
  - Internet sources on all of the above.

NOTE: Sources for this type of information can come from intelligence estimates provided by the higher headquarters' S2/G2, other military units, or the HN.

- **OPSEC information.** This information includes the disposition of US forces (to include MP, MI, CID, and other services' investigating units) deploying or being used within the AO.
- **Other working aids.** These may include items such as computers, printers, boards, and acetate that will be necessary depending on the missions the MP are undertaking.

DETERMINE INFORMATION REQUIREMENTS AND COORDINATE ACCESS TO DATA

E-4. Once the mission analysis is completed, the next step is to determine if there are information gaps. An information gap is a missing piece of information that is critical to the analysis. MP leaders must identify and prioritize the gaps using the time, the available resources, and the commander's intent. Once an information gap is identified, it becomes PCIR. Examples of PCIR are as follows:

- Where is the nearest police station, and how did they handle a certain problem?

- What is the composition of the two gangs operating in our AO, and how can they influence US operations?
- Why is river-crossing point Bravo a danger to US crossing forces? Why are local rioters disrupting only US operations on the east bank?

E-5. The next step is to determine if PCIR can be answered with data already available. To do this, the available military and nonmilitary sources must be identified. Some sources may include—

- **USACIDC.** The USACIDC is responsible for programs such as combating terrorism, CRIMINTEL, personal-security assessment, crime analysis, and LOGSEC threat assessment. Some of the information gathered by the USACIDC (although releasable under existing controls and restraints) may help with the PIAP mission. In fact, USACIDC special agents may already be working in the AO and may be available to help the MP with the collection effort.
- **MI element.** Like the USACIDC, the MI element operating in the AO may have come across a piece of information (police or criminal) that may help accomplish the PIAP mission. The MP must make daily contact with the S2/G2 and coordinate access to invaluable information. As with the USACIDC, available intelligence information may only be releasable under strict controls and restrictions imposed by the commander.
- **Other MP units.** Coordination with an adjacent MP unit may prove to be beneficial, especially if the other MP unit has conducted a similar PIAP mission or has conducted operations in the AOI.
- **HN law-enforcement agencies.** Local law-enforcement agencies will provide the essential populace knowledge otherwise not available to US forces. Chiefs of police, sheriffs, and other key community leaders may help MP leaders fill the information need of the PIAP mission.
- **Joint, interagency, and multinational forces.** Close and effective liaison with these elements will result in quick access to information that can produce the same benefits as the agencies listed above.

NOTE: If, after consulting with all available sources, the PCIR is not resolved, the MP leader must decide the best collection effort to solve the PCIR.

RECOMMEND AND SUPERVISE THE POLICE/CRIMINAL COLLECTION EFFORT

E-6. The collection effort is the means by which specific PCIR will be met. This implies developing a collection strategy, tasking specific collectors, and supervising the collection effort.

- **Develop a collection strategy.** After a through study of the availability, capability, and disposition of the potential collecting resources, MP leaders select which asset is better suitable to perform the mission. Are organic MP the best collectors for this mission? Or do we need to request support from USACIDC units? What unique or organic capabilities do each bring to the effort? Which MP function is the best? Do we conduct MMS, AS, or L&O operations? Part of the collection strategy includes coordinating with the S2/G2, the SJA, the

CID, and other agencies before launching the collection effort. This coordination will eliminate duplication of effort, interference with an ongoing effort, or stepping out of legal limits.

- **Task or request specific collectors.** The missions that will be tasked to respective collectors must be determined. Information collectors can be tasked with more than one mission at a time. However, it is imperative that their tasks be prioritized based on mission requirements and time available. The appropriate tasking or request chain must be used to request an MP team, a platoon, or USACIDC special agents.
- **Supervise the collection effort.** The collectors must be provided with reporting guidelines. How often should they report? Should they report "no information?" Specific reporting instructions should be provided, including how, when, and where they report. A collection tasking chart should also be used (Figure E-2). As more and more collectors are being used, it becomes important to track their missions, capabilities, and success. A brief note on the reporting instructions of each collector should be included.

PIAP Missions	Unit Tasked	Echelon Employed	POC	Collection Strategy Employed	Date Collection Started	Location	Date Collection Ended
PCIR 1							
PCIR 2							
PCIR 3							
PCIR 4							

Figure E-2. Collection Tasking Chart

PROCESS POLICE/CRIMINAL RAW DATA

E-7. This processing step converts raw data into police intelligence through analysis. The key to processing the data is to understand it. For example, as collectors report raw data it may often appear to be meaningless information. Rarely will one collector receive all of the necessary information to answer the PCIR at one time. Instead, a variety of information must be fitted together to form a coherent, even if incomplete, picture.

E-8. Every piece of data is important. This is the cardinal rule of the entire process. Until the data is analyzed for reliability and compared with other data, all reports must be treated equally. This is important to remember, for the trend will be to use data that either fits your expectations or gives the most complete picture.

E-9. A report of "no activity" can be as important as reports with information. If the collector is capable of collecting the information and is actively trying to collect, reports of "no activity" can be critical. First, if there is no indication of a change, they can be used to reprioritize PIAP missions. Second, they can be used to retask the collectors to other areas. It is important to check the mission's original intent periodically to ensure that it is still valid and merits further efforts.

E-10. To determine the reliability of the data, the MP leader must evaluate the source. Does the information come from the police chief, local youth, or other military forces? One of the best ways to determine the reliability of the information is to have more than one collector reporting on the same PCIR. If more than one collector reports the same information, it can normally be presumed to be accurate.

E-11. Information can be analyzed in a variety of ways—chronologically, geographically, by the impact on the force, by the type of activity or the modus operandi, or by the association of participants. The MP leader must evaluate the information after it is analyzed to determine whether the information collected answers the PCIR or whether it helps accomplish the PIAP mission. If the information does not answer these questions, he must decide whether it is necessary to task other collectors, retask the same collectors, or wait until the situation develops.

REPORT AND/OR DISSEMINATE POLICE AND CRIMINAL ASSESSMENT

E-12. Reporting and disseminating is the most important aspect of the PIAP. Collection and analysis is valueless unless the data is communicated to commanders who can best use the information.

E-13. Reporting requirements must be established based on the unit's SOP and other established processes. MP leaders must decide if the information collected must be sent using an established priority format or if it can wait to be sent using SITREPs. Does the information collected affect the force-protection measures or does it immediately impact the tactical scenario? Do we need to alert units operating in the vicinity of AA Fox to avoid sector 2? Additionally, MP leaders must ensure that any information released follows existing constraints and regulations.

E-14. It is also advantageous to provide feedback to the collectors. Collectors often receive no feedback on their performance. Informing them as to which part of their information was vital and beneficial can often increase their productivity in future operations.

SAMPLE SCENARIO

E-15. The following example helps illustrate the implementation of all six steps of the PIAP:

E-16. While deployed in support of Operation Control Chaos, US forces assigned to TF Blue Thunder are tasked with stability and support operations in the city of Corbina. This city, although currently implementing a peace treaty, was the site of extreme civil unrest. Even though the overall situation is not as chaotic as it was before US intervention, there are some sectors that are still in turmoil. Common crimes (such as robberies, assaults, looting, and vandalism) still occur daily and there is fear that this condition may spread out of control throughout the city. The TF commander knows that in order to bring peace to the entire city, he must attempt to stabilize the situation in some of these bad sectors. He decides that the best way to deal with this situation is to assign a sector (sector 51) to an MP battalion in an attempt to control the situation. If the MP battalion succeeds, he may use the same concept throughout the city. He is aware of the five MP functions and how they can help bring this sector under some control. Under this framework, the TF commander issues the following mission: *On order, the MP battalion will conduct stability and support operations in the vicinity of sector 51, the city of Corbina, in support of TF Blue Thunder.* Upon receipt of this mission, the battalion commander gathered his staff and initiated the planning process. Before the MP staff could initiate the PIAP, they made contact with the S2/G2 and received an updated IPB estimate. Now that the IPB estimates are on-hand, the staff initiates the PIAP.

DETERMINE THE SCOPE OF THE PIAP

E-17. The staff performs the following steps to determine the scope of the PIAP:

- **Conduct a mission analysis.** The staff looks at the given mission, the higher echelon commander's intent, and the concept of operations. After considering all of these factors, they produce a restated mission—*On order, the battalion will conduct MP operations in sector 51 to determine the type of threat and to bring this sector under control in support of TF Blue Thunder.*
- **Prioritize PIAP missions.** Since there is only one mission, this subtask is not applicable.
- **Determine the AO and the AOI.** The staff determines that the AO is the entire sector 51 and that the AOIs are the sectors adjacent to 51.
- **Determine key terrain.** The staff determines that the key terrains are the market place, the plaza, and the city park. They select these areas because most of the incidents occurred in these populated areas. Stabilization in these key areas may be advantageous to mission success.
- **Determine required information products/sources.** The end result is the stabilization of sector 51. Minimize the crime incidents and restore L&O. Information should come from the local police chief, local community leaders, civil affairs, PSYOP, and the CID.

ASSEMBLE WORKING AIDS

E-18. The staff assembles the following working aids in support of the PIAP:

- **Maps.** The staff acquires local maps that detail the market place, the plaza, and the city park. Although the maps are in a foreign language, the staff is able to identify these three prominent features and surrounding areas.
- **Language aids.** In addition to dictionaries, the staff is able to get two translators from civil affairs.

- **Open-source information.** Open-source information includes a variety of encyclopedias and country books. Newspaper reports and historical crime data is available via the internet or through MI sources. The emphasis is in obtaining information on the history and composition of the local police force.
- **Classified intelligence.** The staff receives a briefing from the S2/G2 describing the operations of SOF units in the area.
- **Other working aids.** Computers, charts, templates, and acetate are available.

DETERMINE INFORMATION REQUIREMENTS AND COORDINATE ACCESS TO DATA

E-19. After a careful analysis of all available information, the following information gaps were turned into PCIR:

- Why is the police force not actively engaged in L&O?
- What is the major threat to civilians in each area?
- Are local gangs involved?
- Is their local judicial system in place?
- Where are prisoners housed?

E-20. The staff consults with the USACIDC, MI, other MP units, and other sources but determines that the information provided is not clear or sufficient. Since the PCIR are not resolved, the staff decides to recommend that the commander initiate a collection effort.

RECOMMEND AND SUPERVISE THE POLICE/CRIMINAL COLLECTION EFFORT

E-21. The staff performs the following steps to recommend and supervise the police/criminal collection effort:

- **Develop a collection strategy.** Upon coordinating with the S2/G2, the SJA, and the CID, the MP commander decides that the best collectors will come from within organic assets. He decides that he will employ the MP to go out and seek the needed information. The battalion commander will employ a company commander to interview the local police chief and an MP/interpreter will patrol around the designated AOI. The MP commander will need additional interpreters to help him with his collection effort.
- **Task or request specific collectors.** The MP staff requests an interpreter through the normal chain of command. The TF G3 assigns six interpreters for a period of 90 days.
- **Supervise the collection effort.** All collectors are instructed to report not later than (NLT) 1800 every day. The report will be in the form of an "end of day" outbrief to the S2 and the battalion commander. A collection tasking chart (Figure E-3) will also be completed.

PROCESS POLICE/CRIMINAL RAW DATA

E-22. After receiving all of the above information, the S2 concludes that the inability of the police force to do their job has had a domino effect on all areas of peace and order. This information is confirmed through other reports from information collected by the CID (through their own efforts) and the CA. The information also reveals that the gangs have intentions of attacking US forces if they interfere with their "business." The S2 determines that although the situation may seem to be chaotic, once the locals are able to trust the police forces, everything should return to normal.

REPORT AND/OR DISSEMINATE POLICE AND CRIMINAL ASSESSMENT

E-23. Since the collection effort identifies a potential threat to US forces, the commander decides to notify higher headquarters and increase his own force-protection measures. He requires MP elements patrolling in designated areas to be no less than a squad. If patrolling dismounted, a HMMWV or an ASV will trail the squad. The MP commander uses the information provided during the PIAP to develop his implementation plan. This plan includes joint patrols, training, and construction of a joint MP/local national police station.

	PIAP Mission: Bring Sector 51 Under Control							
PCIR	**Unit Tasked**	**Echelon Employ-ed**	**POC**	**Collection Strategy Employed**	**Date Collection Started**	**Location**	**Date Collection Ended**	**PCIR Result**
Why is the police force not actively engaged in L&O operations?	A Co	2 teams	CPT Smith	Direct interview	31 Jan 00	Sector 51, Police HQ	2 Feb 00	The police force does not know how to interact with the population.
What is the major threat to civilians in each area?	A Co	6 teams	1LT Thomas	Patrol and interview	31 Jan 00	Market place, plaza, and park	7 Feb 00	Muggings, robberies, and stabbings are the major threats to civilians.
Are local gangs involved?	A Co	2 teams	CPT Smith	Direct interview	31 Jan 00	Sector 51, Police HQ	2 Feb 00	Two rival gangs exist.
Is a local judicial system in place?	A Co	2 teams	1LT Jones	Direct interview	31 Jan 00	Courthouse	3 Feb 00	A judicial system exists, but the judge is afraid to sentence anyone.
Where are the prisoners housed?	A Co	4 teams	1LT Lucas	AS and MMS	31 Jan 00	Local prison	2 Feb 00	The local prison is abandoned.

SAMPLE

Figure E-3. Sample Collection Tasking Chart

Glossary

1LT first lieutenant

A/L administrative and logistics

AA assembly area

AATF air-assault task force

ABCS Army Battle Command Systems

ACE analysis and control element

ACIRS Automated Criminal Investigative Reporting System

ACT analysis control team

ADA air-defense artillery

ADC area damage control

AFATDS Advanced Field-Artillery Tactical Data System

ALO authorized level of organization

AMDPCS Air- and Missile-Defense Planning and Control System

AO area of operation

AOE Army of Excellence

AOI area of interest

AOR area of responsibility

APOD aerial port of debarkation

APOE aerial port of embarkation

AR	armored
AR	Army regulation
ARFOR	Army forces
AS	area security
ASAS	All-Source Analysis System
ASCC	Army service component command
ASG	area support group
ASP	ammunition supply point
ASR	alternate supply route
ASV	armored security vehicle
ATCCS	Army Tactical Command and Control System
attn	attention
AWIS	Army WMCCS Information System
BCT	brigade combat team
bde	brigade
BFA	battlefield functional area
BLD	brigade liaison detachment
bn	battalion
BOS	battlefield operating system
BRT	brigade reconnaissance troop
BWA	battlefield workload analysis

C^2	command and control
C^3	command, control, and communications
C4I	command, control, communications, computers, and intelligence
C4IFTW	command, control, communications, computers, and intelligence for the warrior
CAS	close air support
CASE	corps-area support element
CCIU	computer-crime investigative unit
cdr	commander
CG	commanding general
CHA	corps holding area
CI	civilian internee
CID	Criminal Investigation Division
CIDR	Criminal Investigation Division Regulation
CINC	commander in chief
CIP	Criminal Intelligence Program
cmd	command
CMOC	civil-military operations center
co	company
COA	course of action
CofS	Chief of Staff
commo	communications

COMMZ	communications zone
conf	confinement
CONUS	continental US
COSCOM	corps support command
CP	command post
CPT	captain
CRC	Crime Records Center
CRIMINTEL	criminal intelligence
CRS-3	Correctional Reporting System
CS	combat support
CSA	corps storage area
CSE	corps support element
CSG	corps support group
CSM	command sergeant major
CSS	combat service support
CSSCS	Combat Service Support Control System
cust	customs
CZ	combat zone
DA	Department of the Army
DC	District of Columbia
DCCP	division central collection point

DCG	deputy commanding general
DIBRS	Defense Incident-Based Reporting System
DISCOM	division support command
div	division
DLEA	domestic law-enforcement agency
DMMC	division materiel management center
DNA	deoxyribonucleic acid
DOD	Department of Defense
DOJ	Department of Justice
DOS	Department of State
DPM	deputy provost marshal
DRA	division rear area
DS	direct support
DSA	division support area
DSB	division support battalion
DSE	division support element
DSS	decisive, shaping, and sustaining
DZ	drop zone
EAC	echelons above corps
EAD	echelons above division
ECB	echelons corps and below

EG escort guard

EMS emergency medical services

EPW enemy prisoner of war

ERP engineer regulation point

ES escort

FA field artillery

FBCB2 Force XXI Battle Command—Brigade and Below

FBI Federal Bureau of Investigation

Feb February

FEBA forward edge of the battle area

FIU field investigative unit

FM field manual

FOB forward operating base

FOIA Freedom of Information Act

FORSCOM US Army Forces Command

FSB forward-support battalion

FSC forward-support company

G1 Assistant Chief of Staff, G1 (Personnel)

G2 Assistant Chief of Staff, G2 (Intelligence)

G3 Assistant Chief of Staff, G3 (Operations and Plans)

G4 Assistant Chief of Staff, G4 (Logistics)

G5 Assistant Chief of Staff, G5 (Civil Affairs)

G6 Assistant Chief of Staff, G6 (Communications)

GAC ground-assault convoy

GCCS Global Command and Control System

GCCS-A Global Command and Control System—Army

GD guard

GO government organization

GS general support

HHB headquarters and headquarters battery

HHC headquarters and headquarters company

HHD headquarters and headquarters detachment

HMMWV high-mobility, multipurpose, wheeled vehicle

HN host nation

HQ headquarters

HRP high-risk personnel

HVA high-value asset

I/R internment and resettlement

IBCT initial/interim brigade combat team

IC information center

IEW intelligence and electronic warfare

IG inspector general

IHO international humanitarian organization

IMETP International Military Education and Training Program

IO international organization

IPB intelligence preparation of the battlefield

IRIC internment/resettlement information center

ISB intermediate staging base

ITO installation transportation officer

J3 Operations Directorate

Jan January

JFC joint-force command

JOA joint-operations area

JP joint publication

JRA joint rear area

JSOA joint special operations area

JTF joint task force

km kilometer(s)

L&O law and order

LID light infantry division

liq liquid

LOC lines of communication

LOGSEC logistics security

LRP long-range plan

LSTA logistics-security threat assessment

LVOSS Light-Vehicle Obscuration Smoke System

LZ landing zone

M/O marksman/observer

M/S mobility/survivability

MACOM major Army command

MAJ major

MCRP Marine Corps reference publication

MCS Maneuver Control System

MDMP military decision-making process

mech mechanized

MEDEVAC medical evacuation

METL mission-essential task list

METT-TC mission, enemy, terrain, troops, time available, and civilian considerations

MEVA mission-essential or vulnerable area

MFPU major fraud procurement unit

MI military intelligence

MILCOM military community

min minimum

MKT mobile kitchen trailer

MLRS	Multiple-Launch Rocket System
MMS	maneuver and mobility support
MOB	main operating base
MOOTW	military operations other than war
MP	military police
MPBTF	military police brigade task force
MPMIS	Military Police Management Information System
MQS	military qualification standards
MSR	main supply route
MTOE	modified table of organization and equipment
MTP	mission training plan
MTT	mobile training team
MWD	military working dog
NATO	North Atlantic Treaty Organization
naut	nautical
NBC	nuclear, biological, and chemical
NCO	noncommissioned officer
NEO	noncombatant evacuation operation
NG	national guard
NGO	nongovernmental organization
NIBRS	National Incident-Based Reporting System

NLT	not later than
NLT	nonlethal weapon
NWP	Naval warfare publication
OCONUS	outside the continental US
OPCON	operational control
OPLAN	operation plan
opns	operations
OPSEC	operations security
ORS-2W	Offense Reporting Systems Windows
PASE	port-area support element
PCIR	police/criminal information requirements
PDF	Panamanian Defense Force
PEO	peace-enforcement operation
PI	police intelligence
PIAP	police information assessment process
PIO	police intelligence operations
PIR	priority intelligence requirements
PKO	peacekeeping operation
plt	platoon
PM	provost marshal
POC	point of contact

POD port of debarkation

POE port of embarkation

POL petroleum, oil, and lubricants

PRC populace and resource control

PSD protective security detail

PSU protective-service unit

PSVA personal-security vulnerability assessment

PSYOP psychological operations

pub publication

PWIS-3 Prisoner-of-War Information System

PZ pickup zone

R&S reconnaissance and surveillance

RAOC rear-area operations center

RC reserve component

ROC rear operations center

ROE rules of engagement

rqmt requirement

RSOI reception, staging, onward movement, and integration

RSTA reconnaissance, surveillance, and target acquisition

RWS remote workstation

S1 Adjutant (US Army)

S2 Intelligence Officer (US Army)

S3 Operations and Training Officer (US Army)

S4 Supply Officer (US Army)

S5 Civil-Affairs Officer (US Army)

S6 Communications Officer (US Army)

SALUTE size, activity, location, unit, time, and equipment

SAW squad automatic weapon

SCI sensitive compartmented information

sec section

SEP separate

SFC sergeant first class

SITREP situation report

SJA staff judge advocate

SMS Security Management System

SOC security-operations cell

SOF special-operations forces

SOFA status of forces agreement

SOI signal operating instructions

SOP standing operating procedures

SPOD seaport of debarkation

SPOE seaport of embarkation

SPOTREP	spot report
spt	support
sq	square
sqd	squad
SRT	special-reaction team
STAMIS	Standard Army Management Information System
stat	statute
STCCS	Strategic Theater Command and Control System
TA	theater Army
TAA	tactical assembly area
TAACOM	Theater Army Area Command
TAC	tactical
TACCIMS	Theater Automated Command and Control Information Management System
TAI	traffic accident investigation
TASE	TSC-area support element
TC-AIMS	Transportation Coordinator Automated Movement Server
TCF	tactical combat force
TCP	traffic-control post
TF	task force
THREATCON	threat condition
TM	technical manual

TMO traffic-management officer

tms teams

TO theater of operations

TOE tables of organization and equipment

TRADOC US Army Training and Doctrine Command

TRP traffic-regulation plan

TSA theater storage area

TSC theater support command

TSE tactical support element

TTP tactics, techniques, and procedures

UMT unit ministry team

UN United Nations

US United States

USACIDC US Army Criminal Investigation Command

USACIL US Army Criminal Investigations Laboratory

USAF US Air Force

USAMPS US Army Military Police School

USC US Code

USCG United States Coast Guard

USCS US Customs Service

USDA US Department of Agriculture

VIP very important person

VRS Vehicle Registration System

WMCCS Worldwide Military Command and Control System

Bibliography

AR 15-6. *Procedures for Investigating Officers and Boards of Officers.* 11 May 1988.

AR 190-12. *Military Police Working Dogs.* 30 September 1993.

AR 190-45. *Law Enforcement Reporting.* 30 September 1988.

AR 190-47. *The Army Corrections System.* 15 August 1996.

AR 195-2. *Criminal Investigation Activities.* 30 October 1985.

AR 525-13. *Antiterrorism / Force Protection (AT / FP): Security of Personnel, Information, and Critical Resources.* 10 September 1998.

CIDR 195-1. *Special Agent Operations Manual.* October 1994.

DA Form 2028. *Recommended Changes to Publications and Blank Forms.* 1 February 1974.

FM 3-0. *Operations.* To be published within the next 6 months.

FM 19-1. *Military Police Support for the AirLand Battle.* 23 May 1988.

FM 19-4. *Military Police Battlefield Circulation Control, Area Security, and Enemy Prisoner of War Operations.* 7 May 1993.

FM 19-10. *The Military Police Law and Order Operations.* 30 September 1987.

FM 19-40. *Enemy Prisoners of War, Civilian Internees and Detained Persons.* 27 February 1976.

FM 34-2. *Collection Management and Synchronization Planning.* 8 March 1994.

FM 34-130. *Intelligence Preparation of the Battlefield.* 8 July 1994.

FM 41-10. *Civil Affairs Operations.* 14 February 2000.

FM 63-4. *Combat Service Support Operations—Theater Army Area Command.* 24 September 1984.

FM 71-3. *The Armored and Mechanized Infantry Brigade.* 8 January 1996.

FM 71-100. *Division Operations.* 28 August 1996.

FM 90-13-1. *Combined Arms Breaching Operations.* 28 February 1991.

FM 90-40. *Multiservice Procedures for the Tactical Employment of Nonlethal Weapons (NLW) (MCRP 3-15.8; NWP 3-07-31; USCG Pub 3-07.31).* 6 October 1998.

FM 100-1. *The Army.* 14 June 1994.

FM 100-5. *Operations.* 14 June 1993.

FM 100-8. *The Army in Multinational Operations.* 24 November 1997.

FM 100-10. *Combat Service Support.* 3 October 1995.

FM 100-15. *Corps Operations.* 29 October 1996.

FM 100-17-3. *Reception, Staging, Onward Movement, and Integration.* 17 March 1999.

FM 101-5. *Staff Organization and Operations.* 31 May 1997.

FM 101-5-1. *Operational Terms and Graphics (MCRP 5-2A).* 30 September 1997.

JP 3-10. *Joint Doctrine for Rear Area Operations.* 28 May 1996.

USC Title 18. *Crimes and Criminal Procedure.* 23 January 2000.

USC Title 32. *National Guard.* 5 January 1999.

Index

Field Manual
No. 3-19.17

Headquarters
Department of the Army
Washington, DC, 6 July 2005

Military Working Dogs

Contents

Distribution Restriction: Approved for public release; distribution is unlimited.

Figures

Tables

Preface

The last Army dog field manual (FM) was published in 1977. It reflected military working dog (MWD) doctrine developed during the Vietnam era. Although useful at that time, much of the information has since become obsolete. Today, MWD teams are employed in dynamic ways never before imagined.

Today's MWD team is a highly deployable capability that commanders have used around the world from Afghanistan to Africa and from the Balkans to Iraq. These specialized teams aid commanders in stability and support operations as well as in warfighting. Being modular and mobile makes these teams very agile. As situations dictate, MWD teams are quick to arrive and able to conduct various operations. Their versatility allows for effective transformation at all echelons among readiness for deployment and operations on the ground, through redeployment and back to readiness.

The highly aggressive dog tactics of the 1960s and 1970s are long gone. Today's MWD program effectively employs expertly trained and motivated handlers coupled with highly intelligent breeds of dogs. These teams are continuously rotating between their assigned duties and deployments worldwide to perform joint operations, multiechelon tasks, and interagency missions.

This FM addresses the current capabilities of the Military Police Working Dog Program as well as the potential for future applications. As technology and world situations change, the MWD team will continue the transformation process and give commanders the full-spectrum capabilities needed to be combat multipliers on the battlefield as well as persuasive force protection and antiterrorism assets.

Appendix A complies with current Army directives, which state that the metric system will be incorporated into all new publications.

This publication applies to the Active Army, the Army National Guard (ARNG)/the Army National Guard of the United States (ARNGUS), and the United States Army Reserve (USAR).

The proponent of this publication is United States Army Training and Doctrine Command (TRADOC). Send comments and recommendations on *Department of the Army (DA) Form 2028 (Recommended Changes to Publications and Blank Forms)* directly to Commandant, United States Army Military Police School, ATTN: ATSJ-DD, 401 MANSCEN Loop, Suite 2060, Fort Leonard Wood, Missouri 65473-8926.

Unless otherwise stated, masculine pronouns do not refer exclusively to men.

Note: For the purposes of this manual, a dog team refers to one handler and one MWD unless specified otherwise.

Chapter 1

The Military Working Dog Program

Dogs have been used for the protection of life and property since ancient times. From these beginnings, dog training and employment has been continuously refined to produce a highly sophisticated and versatile extension of the soldier's own senses. Even the most complex machines remain unable to duplicate the operational effectiveness of a properly trained MWD team.

THE ROLE OF THE MILITARY WORKING DOG

1-1. MWDs provide a valuable asset to military police, infantry, special forces, the Department of Defense (DOD), and other government agencies. The MWD's senses of sight, smell, and hearing enhance his detection capabilities and provide commanders with a physical and psychological deterrent to criminal activity. Properly trained MWDs can prevent an intruder or suspect from escaping. When necessary, the MWD provides an added dimension of physical force as an alternative to the use of deadly force. Public knowledge of MWD team capabilities provides military police and various security forces with a formidable deterrent wherever the MWD team is employed.

1-2. MWDs are key resources for use in the military police combat support role. MWDs are trained for scouting, patrolling, and performing building and area searches. Commanders can use *Table 1-1* to determine what MWD type best suits their needs. Some MWDs have also been trained to track although this is not a required area of expertise. All of these skills contribute to the successful completion of the five military police functions across the full spectrum of military operations.

Table 1-1. MWD Mission-to-Type Matrix

Mission	*MWD Type*		
	PEDD	**PNDD**	**PD**
A. Area Security Operations			
(1) Area and reconnaissance	X	X	X
(2) Screening and surveillance	X	X	X
(3) Base/air-base defense	X	X	X
(4) Cordon and search	X		
(5) Checkpoints	X	X	X
(6) Roadblocks	X	X	X
(7) Response force	X	X	X
(8) Critical site, asset, HRP security	X[1]	X	X
B. Maneuver and Mobility Support Operations			
(1) Maneuver support	X	X	X
(2) Mobility support	X	X	X
C. I/R Operations			
(1) EPW and CI	X[2]	X[2]	X[2]
(2) Evacuation	X	X	X
D. Law and Order Operations			
(1) Force protection	X	X	X
(2) Military police investigations		X	

Table 1-1. MWD Mission-to-Type Matrix (Continued)

Mission	*MWD Type*		
	PEDD	PNDD	PD
(3) Customs support	X	X	
(4) Redeployment operations	X	X	
(5) Suspicious/unattended packages	X[3]		
(6) Health and welfare inspections	X	X	
(7) Crowd control	X[4]	X[4]	X[4]
(8) Alarm responses	X	X	X
(9) Bomb threats	X		
(10) Public MWD demonstrations	X	X	X
E. PIO	X	X	X

[1] PEDDs will not be used to provide security in ASPs, as the ammunition and explosives contained at the ASP will detract from the MWD's detection ability and may distract the MWD from his patrol function.

[2] MWDs will not be used in a correction/detention facility to ensure custody of prisoners.

[3] PEDDs can be used to search the area around a suspicious/unattended package for secondary devices. At no time will a PEDD or handler be used to search the package itself.

[4] Direct confrontation with demonstrators is not recommended, but is authorized with the commander's approval when lesser means of force have been unsuccessful.

1-3. MWDs are a unique item; they are the only living item in the Army supply system. Like other highly specialized equipment, MWDs complement and enhance the capabilities of the military police. MWD teams enable the military police to perform its mission more effectively and with significant savings of manpower, time, and money.

HISTORICAL OVERVIEW OF THE MILITARY WORKING DOG

1-4. Dogs have served in active service at the sides of their handlers for decades. They have been heroes, showing bravery under fire, saving lives (often losing their own), and bringing comfort to the injured and infirmed. The first recorded American use of military dogs was during the Seminole War of 1835 and again in 1842. In Florida and Louisiana, the Army used Cuban bred bloodhounds for tracking. During the US Civil War, dogs were used as messengers, guards, and unit mascots.

1-5. The Army Quartermaster Corps began the US Armed Forces' first war dog training during World War II. By 1945, they had trained almost 10,000 war dogs for the Army, Navy, Marine Corps, and Coast Guard. Fifteen war dog platoons served overseas in World War II. Seven platoons saw service in Europe and eight in the Pacific.

1-6. MWDs were trained at Fort Carson, Colorado, organized into scout dog platoons, and used in the Korean conflict for sentry duty and support of combat patrols. In 1957, MWD training moved to Lackland Air Force base (LAFB), Texas, with the Air Force managing the program.

1-7. Throughout the Vietnam Conflict, the Military Police Corps used dogs with considerable success. Most of these were sentry dogs used to safeguard critical installations such as ports and airfields. A new dimension in canine utilization was realized when marijuana detector dog teams were trained and deployed to assist military police in suppressing illicit drug traffic. Sentry and marijuana detector dog teams were then deployed worldwide in support of military police. An important outgrowth of the conflict was the development of canine research and development efforts. These ongoing efforts were able to initiate the first steps toward developing a more intelligent and stronger military dog, training dogs to detect specific drugs and explosives, developing multiple-purpose dogs, and employing tactical dogs by electronic remote control.

1-8. In the 1990s and early 2000s, MWDs were deployed around the globe in military operations such as Just Cause, Desert Shield and Desert Storm, Uphold Democracy, and Enduring Freedom and Iraqi Freedom. These teams were effectively utilized to enhance the security of critical facilities and areas, as well as bolster force protection and antiterrorism missions, allowing commanders to use military police soldiers and other assets more effectively elsewhere. MWD operations in Haiti showed the first large-scale utilization of MWDs since Vietnam. In the Balkans, military police deployed with MWDs to provide force protection and antiterrorism support and area security throughout the US base camps. MWDs trained as explosive detector dogs conducted vehicle sweeps and building searches to ensure the safety of US soldiers. US Army MWD teams operated beside Iraqi police officials to help in law and order missions and restore the war torn country into a peaceful region.

1-9. MWDs are force multipliers. Installation commanders should include MWDs when planning for force protection and antiterrorism countermeasures. They expand the individual soldier's effectiveness in the face of those that would come against him. They provide an immense physical and psychological effect when used as a show of force in day-to-day operations. The various uses of MWDs have been effectively employed in many aspects of military police missions. MWDs are utilized effectively at gates, camps and bases, and checkpoints and for random searches for narcotics and explosive devices. MWDs are also utilized for other missions in support of combat, combat support, and combat service support units.

1-10. MWD teams are assigned to perform high-risk personnel (HRP) security missions. MWD teams perform narcotic detection missions on the Mexico and US border to assist US Customs. They have been deployed on the United States and Canada border to perform law and order operations. The canine can easily detect what his handler cannot see, hear, or smell. The MWD's controllable aggressive behavior, coupled with the physical and psychological effects he creates, makes him and his handler an impressive and unmatchable team.

ORGANIZATION AND STRUCTURE OF MILITARY WORKING DOG UNITS

1-11. The US Army MWD program is an organization with structure and responsibilities that reach from the DOD level down to the handler level. For information concerning organization and responsibilities above the local level, see *Army Regulation (AR) 190-12*.

Kennel Master Team

1-12. The kennel master (KM) team is comprised of two noncommissioned officers (NCOs) (one sergeant first class [SFC], who is the KM and one staff sergeant [SSG], who is the plans NCO). The KM team is modular and provides technical supervision of MWD teams assigned to the military police brigade. This team may be assigned to the corps military police brigade, echelon above corps (EAC) military police brigade, or the internment and resettlement (I/R) brigade. In addition to technical supervision, this team plans and coordinates MWD operations and support requirements. The KM team is also responsible for MWD team proficiency training. This team supervises the storage and accountability of the narcotic and explosive training aids as well as all assigned equipment.

Patrol Explosive Detector Dog Team

1-13. The patrol explosive detector dog (PEDD) team is comprised of three NCO MWD handlers (one SSG, and two sergeants [SGTs]) and three PEDDs. This team detects explosives and contraband in support of combat, combat support, and combat service support units and the five military police functions. It provides MWD patrol support (including I/R operations) when not employed with explosive detection. This team is normally assigned to the corps military police brigade, EAC military police brigade, military police I/R brigade, or the I/R command and control (C2) detachment. This entire modular team can be deployed together as a unit with all assigned equipment or, based on the specific mission, a handler and an assigned MWD may be deployed apart from the team.

Patrol Narcotic Detector Dog Team

1-14. The patrol narcotic detector dog (PNDD) team is comprised of three NCO MWD handlers (one SGT and two specialists [SPC]) and three PNDDs. This team detects narcotics and other drugs in support of combat, combat support, and combat service support units and the five military police functions. It provides MWD patrol support (including I/R operations) when not employed with narcotics detection. This team is normally assigned to the corps military police brigade, EAC military police brigade, military police I/R brigade, or the I/R C2 detachment. This entire modular team can be deployed together as a unit with all assigned equipment or, based on the specific mission, a handler and an assigned MWD can be deployed apart from the team.

INDIVIDUAL DUTIES AND RESPONSIBILITIES

1-15. There are five levels of responsibility in the MWD program. The terminal level is the program manager and the beginning level, the handler. This is not to be confused with the operational chain of command or standards of grade. These individuals have specific duties in managing and performing the MWD program.

PROGRAM MANAGER

1-16. The program manager executes major Army command (MACOM) level responsibilities for the MWD program. He is the equivalent of an operations NCO and is an appointed certification authority whose duties include—

- Conducting MWD team patrol and detector certifications.
- Requisitioning personnel and MWDs.
- Providing policy and guidelines for the MACOM.
- Forecasting annual operational budgets.
- Coordinating MWD support for the office of the Secretary of Defense, US Secret Service (USSS), US Customs Service (USCS), and Department of State missions.
- Serving as the MACOM provost marshal's (PM's) advisor for the employment of MWDs in contingency operations, force protection and antiterrorism support, and homeland defense.

KENNEL MASTER

1-17. The KM is the senior NCO in charge of the installation's MWD program, consisting of 6 to 30 handlers and respective MWDs. KM duties are the equivalent of a platoon sergeant's duties when 9 or more handlers and MWDs are assigned. When 8 or fewer handlers and MWDs are assigned, the duties are equal to those of a squad leader. In addition to traditional leadership responsibilities, KM duties include—

- Monitoring all training to ensure that MWD teams are ready for annual certifications and worldwide deployments in support of various operational missions.
- Advising the commander on the employment of MWD teams.
- Ensuring that team proficiency is maintained.
- Preparing for MACOM and veterinary inspections.
- Forecasting annual operational budgets.
- Assisting the MACOM program manager in requisitioning personnel, MWDs, and equipment.
- Managing the daily maintenance and upkeep of kennel facilities.
- Monitoring MWDs' health and welfare.
- Managing procurements and accounts for explosive and narcotic training aids.
- Advising installation and unit force protection and antiterrorism officers on the use of MWDs.
- Coordinating MWD support for the USSS, the USCS, DOD protective services units, combat support operations, the Army Recruiting Command, and other civilian agencies.

MILITARY WORKING DOG PLANS NONCOMMISSIONED OFFICER

1-18. The MWD plans NCO is part of the KM team and performs the duties of the maneuver mobility support operations (MMSO) and training NCOs. The plans NCO is selected for his ability and overall knowledge of training, mentoring, and deploying MWD assets. This NCO is responsible for—

- Developing training plans.
- Coordinating and planning the unit mission-essential task list (METL).
- Supervising the execution and tracking of the deployment plan.
- Tracking the movement of MWD teams while they are deployed.
- Ensuring that deployed soldier's issues are resolved.
- Assisting in the supervision of the installation MWD program.
- Acting as the KM when the KM is absent.
- Preparing short- and long-term training outlines and risk assessments.
- Validating and supervising proficiency training.
- Overwatching preparation for the deployment of MWD teams.

SENIOR MILITARY WORKING DOG HANDLER

1-19. The senior MWD handler is the senior NCO of the PEDD or PNDD teams. His duties are the equivalent of the squad leader of the PEDD team or the team leader of the PNDD team. In addition to traditional leadership responsibilities, he is responsible for ensuring that—

- Annual certification training is conducted.
- MWD teams are ready for deployments.
- Team proficiency is maintained.
- Handlers are proficient with their assigned MWDs.
- Coordination for MWD team support for the USSS, the USCS, combat support operations, the Army Recruiting Command, and other civilian agencies is completed.

MILITARY WORKING DOG HANDLER

1-20. The handler provides the daily care and grooming for his assigned MWD. He ensures that proficiency skills are maintained in his assigned MWD. The handler and MWD perform either PNDD operations or PEDD operations. The handler and MWD are all trained for patrol but the variant for narcotics detector or explosive detector depends on the MWD team's training. The handler also—

- Fills out training records daily.
- Grooms the assigned MWD daily.
- Trains the assigned MWD daily.
- Maintains MWD kennels daily.
- Performs additional kennels duties.

Note: For more information about additional skill identifier (ASI) Z6 standards of grade, refer to *DA Pamphlet (Pam) 611-21.*

FUNDAMENTALS FOR THE USE OF MILITARY WORKING DOGS

1-21. An MWD can be trained to respond consistently to certain sensory stimuli (sights, sounds, odors, or scents) and alert the handler. Under most circumstances, a properly trained MWD can smell, hear, and visually detect movement better than a person.

1-22. Commanders and supervisors should consider the capabilities of the MWD team when identifying and evaluating possible MWD employment. Not all MWD teams are able or equipped to perform all missions equally (see *Table 1-1,* page 1-1).

USE OF PATROL DOGS

1-23. All MWD teams are trained to perform patrol missions. One benefit of employment is the deterrence of criminal activity. The obvious presence and well-published activities of MWD teams can successfully deter trespassers, vandals, violent persons, and other would-be criminals. When the desired task is to deter unauthorized or suspicious individuals, an MWD team should be assigned to a location during day or night.

1-24. The patrol dog's (PD's) contribution to the law and order effort is most effective when the team is utilized as a walking patrol. The law enforcement duties that a PD team can perform as a walking patrol include checking or clearing buildings and patrolling parking lots, family housing areas, and troop billet areas. Mobility significantly increases the potential area of coverage. The best and most effective method is the ride awhile, walk awhile method. This concept helps cover large patrol areas, and the periodic exercise helps keep the team alert. All PD teams can be utilized for—

- Antiterrorism operations.
- Force protection and antiterrorism efforts.
- Special-reaction team (SRT) operations.
- Assistance in military police investigations.
- Walking and mobile patrols.
- Alarm responses.
- Building checks and searches.
- The identification and apprehension of individuals.
- The protection of HRPs.
- The protection of funds.
- Open-area searches.
- Civil disturbances.
- Perimeter security.
- External intrusion detection.
- Response forces.
- Access control points (ACPs) support.
- High-risk targets (HRT) support.
- Mission-essential vulnerable areas (MEVA) support.
- I/R operations.

1-25. Many of the tasks above can be performed in combat support or force protection and antiterrorism operations. In addition, PD teams can and should be utilized as a tool for area defense, providing warning and response. PD teams are effective on combat patrols and as listening posts and reconnaissance team members. The employment of PD teams should not be overlooked during I/R or enemy prisoner of war (EPW) operations. PDs are trained to apprehend suspects at or near a crime scene, stop those who may attempt to escape, and protect their handlers from harm. Some PDs may be able to locate suspects who have left the scene of a crime.

1-26. The PD team should be utilized to provide a psychological deterrent and a show of additional force protection and antiterrorism measures. PD teams should be placed on roving patrols at ACPs, vulnerable areas, and the location of HRP. These missions should increase as force protection and antiterrorism levels increase.

USE OF PATROL NARCOTIC DETECTOR DOGS

1-27. Military police commanders, planners, and supervisors have a very effective means of detecting the possession and/or use of controlled substances with the employment of a PNDD team. PNDDs are trained to recognize the scent of certain illegal substances through a program of practice and reward.

1-28. In addition to PD missions, PNDD dogs can be utilized for the following:

- Criminal Investigation Division (CID) and military police investigator drug investigations.
- USCS inspections.
- Health and welfare inspections.
- Barracks searches.
- Workplace searches.
- Random gate inspections.
- Area searches.
- Vehicle searches.
- Postal inspections.
- I/R operations.
- Aircraft and luggage searches.
- Predeployment and redeployment searches.
- Other US government agency support.

1-29. When the PNDD responds on any of the substances he is trained to detect, his handler takes the appropriate action. PNDD teams perform a valuable service by helping to rid the military community of illegal drugs and the problems associated with drug and controlled substance abuse.

USE OF PATROL EXPLOSIVE DETECTOR DOGS

1-30. PEDD teams are very useful in many searches or investigations involving explosives. The search for explosives is critically important to ensure that the community is safe and protected from would-be terrorists. PEDD teams are particularly useful if there is a need to locate one or more of these explosive items, which may have been hidden in an area.

1-31. The deterrent value and the detection capabilities of PEDD teams are a very effective countermeasure to terrorism. Public knowledge that PEDD teams are in use is a deterrent to persons who may try to use explosives illegally. The knowledge that explosives can be detected by PEDD teams at entry points or in places where explosives have been hidden can prevent a person from attempting to commit terrorist acts.

1-32. In addition to PD missions, PEDD dogs can be utilized for—

- Bomb threat incidents.
- Suspicious/unattended package incidents.
- ACP searches.
- Checkpoint searches.
- Random gate searches.
- CID and military police investigations section.
- Health and welfare inspections.
- Building and area searches.
- Postal inspections.
- Aircraft and luggage searches.
- Predeployment and redeployment searches.
- Protective service missions.
- Other US government agency support.

1-33. A well–trained PEDD team can conduct a significantly more effective search of an area or facility in a much shorter time than a number of people can. Using PEDD teams helps reduce the potential risk to persons who would otherwise have to do the search without the benefit of the dog's superior sense of smell. More detailed information concerning the employment of PD, PNDD, and PEDD teams is discussed in *Chapters 5, 6,* and *7.*

Chapter 2
Administrative Requirements

This chapter contains information that is useful in planning an MWD program or adding MWD teams to an existing program. The guidelines in this section are not meant to be absolutes for utilization, training, or maintenance. They give general guidance and information that should be considered when planning a program. Leaders will find the following information helpful for writing authorization documentation statements and for planning the local MWD program. It is important to remember that all MWDs are trained as PDs in addition to the specific type of detector dog (narcotic or explosive).

DETERMINATION OF A NEED FOR MILITARY WORKING DOGS

2-1. The installation provost marshal (PM), military police unit commander, or task force commander must take the initiative to establish the local MWD program if it does not already exist. The specific needs must be determined and the costs of the program must be justified. The decision process involves a thorough risk and crime analysis, an accurate evaluation of the requirements of the entire military police mission, and the use of the working dog yardsticks located in this chapter.

2-2. Some of the factors to consider when determining the need for MWD teams include—

- The mission of the team.
- The size of the area of responsibility of the team.
- The size of the installation or the area population.
- The number of personnel to be served by the team.
- Incident rates for all crimes.
- The present capability and the commitment of a portion of the manpower resources as handlers.
- The types of terrain on the installation or in the probable areas of deployment.
- The types of combat support missions for which MWD teams can be used.
- The number of critical facilities or areas.
- The capabilities of each type of team.
- The number of HRP on the installation or in the area of responsibility.

VERIFICATION OF AUTHORIZATIONS

2-3. Initial MWD authorizations should be probationary using the following criteria:

- Adjustments must be made based on actual use data.
- A baseline ratio of 24 hours of MWD utilization to 8 hours training per week (4 hours patrol and 4 hours detection) must be maintained.
  - Utilization includes all missions (force protection/law enforcement, security, and combat support), whether patrolling, detecting drugs or explosives, or performing other functions (in combination), when a handler and dog are being employed together as a team.
  - Training includes all training activities to maintain, improve, regain, or develop dog skills such as obedience, controlled aggression, patrolling, scouting, detection of persons, and detection of drugs or explosives.
- An automatic review must be made by the respective MACOM after two years.

2-4. There is a trade-off when adjusting between the training and utilization necessary to reinforce basic skills or add new skills. For example, four hours per week may be adequate to maintain detector dog proficiency or to train in new skills, but additional training may be justified at times and training hours may be increased accordingly. This increase in training time should be accompanied by an equivalent reduction in utilization time. However, training hours should not be increased merely as a substitute for lack of utilization.

MISSION REQUIREMENTS

2-5. The mission requirements for the following support missions should be considered. These requirements should be used only as guidelines.

INSTALLATION FORCE PROTECTION AND LAW ENFORCEMENT SUPPORT

2-6. In order to provide continuous 24-hour, year-round MWD protection of the installation, nine MWDs are needed. Six of these dogs will be PEDD and three will be PNDD. This requirement is based on each MWD team being used a minimum of 24 hours per week. This does not include hours needed for training, grooming, or health care requirements. This does allow for temporary duty (TDY), handler leave, and illness.

2-7. The goal is to have one PEDD per installation available at all times for bomb threats or explosive type missions (not counted as part of the working force protection requirement). Installations with a three- or four-star headquarters require one additional PEDD authorization. This is due to the increased workload, facility criticality, facility vulnerability, location, past history of terrorist target selections, and present threat assessments.

HIGH-RISK PERSONNEL SECURITY MISSIONS

2-8. DOD tasks the Army to provide PEDD support for presidential, vice presidential, Secretary of State, and Secretary of Defense visits throughout the world. Currently, there is not a published formula to predetermine mission frequencies or requirements for these missions. On average, 40 teams inside the MACOM support the various missions above. These figures do not include travel or mission preparation time.

CUSTOMS SUPPORT

2-9. While no formula exists to quantify requirements for military customs support, this remains an ongoing mission that must be viewed as a requirement for additional MWD authorizations. MWDs are used to screen household goods shipments, privately owned vehicles (POVs), container express (CONEX) containers, military equipment, US mail, and port operations. Workload varies by installation population, location, unit missions and deployments, and the time of year.

CONTINGENCY OPERATIONS SUPPORT

2-10. Currently, approximately 50 MWD teams are supporting various contingency operations. This requirement will continue until mission completion. This figure represents 25 percent of Army MWD authorizations. Support for contingency operations has continued to rise since 1990 and will continue to do so. This increase must be factored into planning for MWD authorizations. A conservative average of 100 MWDs is needed for contingency operations on a yearly basis. This number is to ensure that MWD teams do not perform back-to-back deployments and that they receive proper proficiency training according to *AR 190-12* (4 hours per week in each specialty and 95 percent accuracy for PEDDs or 90 percent accuracy for PNDDs).

KENNEL CONSTRUCTION

2-11. When new MWD programs are started at locations where there are no facilities in place, kennel construction must be complete before the dogs arrive. The MWDs should be ordered through the military standard transportation and issue procedures (MILSTRIP) process approximately one year before the expected completion date of the kennel facility. In operations where MWD teams will be deployed for long periods of time and will be rotated, a permanent facility must be constructed within one year. Deployable field kennels may be used in the interim. The joint service MWD committee states that the DOD's design guide for MWD facilities must be referenced in planning the construction or renovation of a kennel facility.

2-12. Once the need and the justification for an MWD program are established, the PM, military police commander, or task force commander initiates a request to the installation facilities engineer to develop the kennel facility design. Before the kennel design has been finalized, the activity or installation veterinary corps officer (VCO), the physical security inspector, safety personnel, and the KM must be consulted to ensure that the minimum requirements are included in the final design. After kennel designs are completed and construction costs are estimated, the kennel project is submitted to the installation commander for approval and funding. The request for authorization of MWD teams must include the statement that the commitment has been made to fund and build kennels within one year prior to the arrival of the MWDs at the installation. Refer to *Appendix B* and *DA Pam 190-12* for details on kennel construction.

PROCUREMENT

2-13. All MWDs must be procured from the procurement section at the 341st Training Squadron, Lackland AFB, Texas. Local procurement of an MWD is prohibited. Requisition of an MWD team (one dog and one handler) requires two separate transactions. Handlers are obtained as a result of a personnel action requesting military police soldiers with the ASI Z6 in the required pay grades. MWDs are obtained as a result of a supply action. Commanders will assign an appropriately qualified handler to every MWD on hand. Assignment will be consistent with the policy of "one dog and one handler" as outlined in *AR 190-12*.

2-14. When the replacement of an MWD and not a handler is needed, requisitions are done according to *AR 700-81*. Initial requisitions for dogs should be submitted when positions for handlers are authorized. The MWDs in the Army supply system are—

- Dog, patrol: narcotic detector; national stock number (NSN) 8820-00-435-7542, line item number (LIN) D33800.
- Dog, patrol: explosives detector; NSN 8820-00-188-3880, LIN G33732.

2-15. The method of how to replace an MWD that has died, been adopted, or has undergone euthanasia is to submit a requisition in the MILSTRIP format that includes a statement that the unit or installation already has a trained handler on station. The MILSTRIP should also be submitted when the VCO has stated in writing that an MWD only has a 12-month life or work expectancy and/or is placed in deployment category IV (permanently nondeployable).

2-16. KMs should send requisitions for new or replacement dogs, in the MILSTRIP format, to their local property book office and forward a copy to their respective MACOM. The headquarters of the Air Force Security Police Agency will review all requisitions for dogs prior to processing by the 341st Training Squadron, Procurement Section.

TRAINING REQUIREMENTS

2-17. Training is needed at each installation to ensure that teams remain proficient. This realistic and challenging training should be scenario driven and relate to all of the possible operations in which MWD teams will operate according to the unit METL. Properly trained MWD teams are better prepared to accomplish all facets of MWD missions.

Initial Training

2-18. MWD teams are initially trained at the 341st Training Squadron, Lackland AFB, Texas. This training is done at a basic level for both the handler and the dog and is not intended to prepare either the handler or the dog for immediate duty in an MWD team. The training is only intended to give a basic overview of the MWD program, the health and care of MWDs, and proper training techniques. It is the responsibility of the gaining unit to train a new handler or dog on all missions conducted by Army MWD teams. Proficiency training is continuously required to ensure that teams are reliable for use in patrol, narcotic detection, and explosive detection.

Proficiency Training

2-19. The effectiveness of MWDs trained in patrol, narcotic detection, or explosive detection depends on the continual reinforcement of the abilities of the dog team through proficiency training. This training is mandatory and if it is not conducted properly, an appointed certification authority can decertify the MWD team. Failure to meet the minimum training standard for at least the 35 days prior to a certification action will result in MWD teams not being evaluated for certification.

2-20. The minimum training standard is at least four hours of training each week in patrol and at least four hours of training each week in narcotic or explosive detection. The minimum standard of proficiency to maintain certification as a detector team is 90 percent or higher (with no more than a 10 percent false response rate) for narcotic detector dogs, and 95 percent or higher (with no more than a 10 percent false response rate) for explosive detector dogs.

2-21. Failure to maintain an average that meets or exceeds the minimum standard for three consecutive training events will result in the automatic decertification of the MWD team. Decertified teams may not be used in any way, and must undergo intensive retraining. These teams may be recertified only after retraining and a consistent demonstration to an appointed certification authority. Proficiency training should be conducted both on and off leash.

Proficiency Training of Patrol Dog Teams

2-22. The effectiveness of PD teams depends on the continual reinforcement of mandatory proficiency training. PD teams must receive a minimum of 4 hours of patrol proficiency training each week. The training environment should approximate the military police working environment as closely as possible. Training should occur at varying times throughout the day and night and on varying days of the week, including weekends.

2-23. The emphasis in training is to develop the skills of both the handler and the dog so that they complement each other and the team becomes a working unit. As a proficiency evaluation criterion, the correct performance of the required task verifies that the handler understands how to control his PD and is able to do so.

2-24. Train the PD to protect with or without command based on any perceived threat to the handler. Because of the potential of serious bodily injury in this specialty, the certification of the handler and the dog as a team is very stringent. Ensure that the proficiency training of PDs includes an emphasis on critical tasks. There are no waivers for failure to accomplish a critical task by a PD team.

Weapon Fire Training

2-25. Training on weapon firing ranges is essential for the PD to become proficient and not be deterred from attacking agitators during gunfire. The dog must not attack the handler during gunfire. The firing of all weapons assigned to the handler should be done with the MWD present whenever possible. PDs can be desensitized with the firing of many different types of weapons. This can often be accomplished by arranging for the handlers to take the dogs to weapon ranges of different units.

2-26. Determine the dog's reaction to the sound of gunfire. It may be necessary for the handler to use counterconditioning techniques until the desired proficiency is achieved. Counterconditioning techniques include starting at distances of 300 meters and slowly bringing the gunfire closer to the dog (or as safety allows, bringing the dog closer to the weapon). The goal is for the dog not to bark or show any signs of aggression when the handler fires all assigned weapons. This can be a slow process over the course of several days.

Obedience Course Training

2-27. The obedience course exposes the PD to various obstacles (see *Figure 2-1*) that simulate walls, open windows, tunnels, ramps, or steps. The dog's exposure to these obstacles reduces the amount of time required to adapt dogs to different environments. The dog learns to negotiate each of the obstacles. When confronted with a similar obstacle in the working environment, the dog is not deterred from completing his mission. The obedience course also develops the handler's ability to control the dog's behavior both on and off leash. The obedience course is not a substitute for exercise. A dog should never be required to negotiate the obedience course until he has been warmed up by proper exercise.

Figure 2-1. Obedience Course Obstacles

Proficiency Training of Detector Dog Teams

2-28. Continuous training of detector dogs is required so they maintain proficiency in the ability to seek and find drugs or explosives, depending on their specialty. Ensure that the detector training is always documented properly. Documentation and training are critical to both maintaining team certification and to establishing probable cause for a search based on a detector dog's number of correct responses.

2-29. There are several differences in the training requirements for PNDD and PEDD. Proficiency training scenarios for both types of detector dogs must be varied to avoid conditioning the dogs to a repetitious training pattern. The following factors will be varied for each proficiency training exercise:

- The general training exercise area.
- The number of training aids planted.
- The specific locations of the training aids in the area being used for training.
- The amount of explosives or drugs used in the training aid plant.
- The type of explosives or drugs used.
- The time of day or night of the training.
- The type of training aid container.
- The type of distraction material planted with the training aid.
- The length of time the training aid is left in place before the detector dog team search.
- The person used to handle and/or plant the training aid.
- The height above or below the floor level of the training aid plants.
- The size of the room or area in which the training aids are planted.

DECOYS

2-30. The decoy is the person who role-plays the primary adversary for training and evaluating the MWD team. The decoy may be a suspect, a subject, an attacker, an agitator, a drunk, an escapee, an enemy, or any of a number of other persons an MWD team may expect to encounter while performing military police duties. A decoy may also be neutral or an ally, such as another military policeman, a supervisor, a lost juvenile, or an innocent person passing through an identification check. Use both males and females for decoys, agitators, and suspect role-players. Vary the clothing the role-players are dressed in to expose the dogs to civilian, military, and the ethnic dress of different countries, if possible.

2-31. Decoys must be taught how to handle the arm protector sleeve or wrap and other aggression tools such as the bite suit (*Figure 2-2*), the hidden wrap, and leg wraps for self–protection when the dog attacks. The sleeve or wrap used should be a hidden sleeve rather than the full, heavy padded sleeve. The hidden sleeve can be concealed under a sweater or field jacket to achieve more realistic conditions for training the MWD.

Figure 2-2. Bite Suit

2-32. KMs and plans NCOs teach decoys how to understand and use the different drives the MWD shows. Once understood, these drives or needs of the MWD can be exploited to cause the desired result in training. Decoys must learn the type of pattern to take to and from the MWD when training on aggression tasks or the bite. For safety and optimum training effect, decoys will learn how to target the MWD into the bite gear and what actions to take during a partial and full bite.

VALIDATION

2-33. The KM or plans NCO will conduct a validation prior to requesting certification. Validations will then be conducted quarterly to verify the patrol and detection accuracy rates recorded on *DA Form 2807-R (Military Working Dog Training and Utilization Record)* and/or *DA Form 3992–(R) (Narcotics or Explosives Detector Dog Training and Utilization Record)*. Additional validations will be given to any patrol or detector MWD whose demonstrated proficiency appears to differ significantly from the rating recorded on his *DA Form 2807-R* and/or *DA Form 3992–R*. The validation will verify proficiency or identify the specific patrol/detector weaknesses that need corrective training.

2-34. Validations include at least 30 trials for a narcotic detector dog and at least 40 trials for an explosive detector dog. These trials will be divided equally with each substance the dog is trained to detect. Training aids will be distributed in at least five different areas over a five–day period. These areas should include (if available) vehicles, barracks, open areas, warehouses, theaters, planes, and luggage. A written record of the evaluation and validation will be maintained on record for a minimum of two years.

CERTIFICATION OF MILITARY WORKING DOG TEAMS

2-35. *AR 190-12* defines the standards and specific requirements for all certification of MWD teams. Certification of an MWD and handler is valid for one year and is immediately nullified when—

- The dog is assigned to another handler.
- The handler and MWD have been separated or have not conducted training for thirty-five or more consecutive days.
- The team fails to meet the minimum detector standard of 95 percent accuracy (explosives) and 90 percent accuracy (narcotics) with no more than a 10 percent false response rate for three consecutive months.

SCOUTING OR PATROLLING

2-36. The PD must be able to detect a person by scent, sound, and sight (on leash only). This task is rated critical. The following tasks must be accomplished to demonstrate proficiency in scouting or patrolling:

- Daytime standards.
  - Detect and respond to the scent of a person who is hidden within 75 yards, and follow the scent to the person's location.
  - Detect and respond to a sound made by a person 100 yards away, and follow this sound to the source.
  - Visually detect and respond to a person in view 100 yards away, and pursue the person on command.
  - Demonstrate proficiency in pursuit, attack, and release.
- Nighttime standards:
  - Detect and respond to the scent of a person who is hidden within 75 yards, and follow the scent to the person's location.
  - Detect and respond to a sound made by a person 100 feet away, and follow this sound to the source.
  - Demonstrate proficiency in pursuit, attack, and release.

Note. There is no nighttime standard for sight due to the dog's natural ability to revert to other senses in the dark.

VEHICLE PATROL

2-37. The PD must be able to ride quietly and calmly inside a vehicle with the handler until provoked or commanded. This is a semicritical task.

TRACKING

2-38. Tracking is the only noncritical task on which PDs are trained. Tracking is noncritical because not all PDs have tracking potential. The local KM should know which dogs have been trained to track. PDs that demonstrate a potential for tracking are identified by the KM or plans NCO at each unit, and the dogs are then further developed in this task. Although tracking is a valuable skill for the PD team, improper performance of the task does not degrade the overall effectiveness of the PD team to perform the military police mission.

2-39. PDs selected for tracking training should be trained and employed for tracking on a tracking harness. The harness serves as a useful cue to the PD that it should focus his senses on tracking. At the start of the track, the handler will give the command "Track" and begin casting for the scent. Initially, scent pads are identified and then an item with the tracklayer's scent on it will be used. The dog is required to track and alert on any items left by the tracklayer.

2-40. The minimum level of proficiency a PD trained for tracking must attain is to be able to follow a human scent at least one hour old, for 1 mile over varied terrain, and on a course with several turns. Given suitable tracking conditions, a skilled tracker dog team can follow the natural wanderings of individuals or groups of persons for at least 3.1 miles over rugged and varied terrain and on a scent track that is at least 12 hours old. While tracking, the dog is also capable of alerting his handler to the presence of tripwires and ambushes.

2-41. Evaluation of tracking tasks must include various ages and lengths of the track. The areas used to conduct tracking training and evaluations will be rotated frequently. Tracks will be at least one-third of a mile long but no longer than one mile. The age should vary from ten minutes to one hour. Tracks will include turns, but evaluators should ensure that the person laying the tracks is not trying to deceive the dog with undue tracklaying. Articles will be placed on the track for the dog to find.

INSPECTIONS

2-42. Inspections help ensure that the MWD section is in compliance with Army standards. *Appendix C* contains an example inspection memorandum to assist the inspection officials on what areas to inspect at an MWD section.

DEPARTMENT OF THE ARMY INSPECTIONS

2-43. Personnel from headquarters, Department of the Army (HQDA), conduct annual inspections of the MACOM MWD program. These inspections are used to assist the MACOM PM and program manager in moving the MACOM MWD program forward. This inspection will be on all aspects of the MACOM MWD program (see *Appendix C*).

MAJOR ARMY COMMAND INSPECTIONS

2-44. MACOM program managers conduct annual inspections of handlers, MWDs, training, team use, team proficiency, equipment, and kennel facilities. This ensures compliance with *AR 190-12*.

PROVOST MARSHAL AND COMMANDER INSPECTIONS

2-45. PMs and/or commanders conduct monthly inspections of handlers, MWDs, training, team use, team proficiency, equipment, and kennel facilities. These inspections are conducted at least monthly on one or more areas of the kennels and quarterly on all aspects of the installation MWD section to ensure compliance with *AR 190-12*.

2-46. A written record of the monthly and quarterly inspection is maintained on file with the appropriate corrective action (both in memorandum format), stating the corrective action taken for each deficiency. This written record is maintained for one year past the inspection date.

2-47. The responsible VCO will conduct an inspection of the kennel facility at least quarterly to ensure health and welfare for both the handlers and dogs assigned to the MWD section. A copy of this inspection is maintained on file for a minimum of two years. The responsible PM and/or commander takes all actions necessary to correct any deficiencies noted by the VCO. Veterinary inspections are to be conducted monthly, if possible.

TRAINING RECORDS

2-48. There are two forms that must be used to annotate MWD training, *DA Forms 2807-R* and *3992-R*. These forms are official military documents and must be maintained daily with factual information. Many times these documents are used as court documents, so the importance of the completion and accuracy of these documents cannot be stressed enough. Example forms and the procedures for filling these forms out are in *Appendix D*.

PROBABLE CAUSE FOLDERS

2-49. Probable cause folders are one of the most important methods of proving an MWD team's reliability. Legal requirements to conduct a search of private property based on the response of an MWD do not allow the probable cause folder a margin for error. All the items contained in the probable cause folder are copies. The following original records must be kept on file at the kennel facility:

- ***Department of Defense (DD) Form 1834 (Military Working Dog Service Record).*** *DD Form 1834* must be maintained on the MWD and kept as the initial form in the probable cause folder. This form must be accurate and signed on a change of a handler. When the MWD is trained on new substances, ensure that they are listed on the *DD Form 1834*. The 341st Training Squadron at Lackland AFB, Texas, must approve new substances.
- ***Lackland AFB Form 375 (Patrol Dog Certification).*** *Lackland AFB Form 375* follows the *DD Form 1834* in the probable cause folder and lists patrol certification information for the MWD.
- ***Lackland AFB Form 375a (Detector Dog Certification).*** *Lackland AFB Form 375a* follows the *Lackland AFB 375* (if dual certified), and lists postgraduation certification information for the MWD.
- ***Air Force (AF) Form 1256 (Certificate of Training).*** A copy of the handler's *AF Form 1256* follows *Lackland AFB Form 375a*. The handler provides a copy of the *AF Form 1256* for the probably cause folder.
- **Certification Letter**. This letter attests to the fact that the appointed certification authority has witnessed a demonstration of the capabilities of the dog team. This letter, at a minimum, should list the handler and dog, the whelp, the type of MWD (PEDD or PNDD), the social security number (SSN) of the handler, and the date of certification. This letter should be maintained in the probable cause folder after the handler's graduation certificate until the next certification of the team or a change of handler.
- **Quarterly Validation and Evaluation by the KM**. This document is used to show that the KM has conducted a quarterly evaluation on the MWD team. If the KM is not available, the plans NCO can also conduct this evaluation. This document follows the certification letter in the probable cause folder.
- **Training and Utilization Records**. The last items that should be maintained in the probable cause folder are the last three months of training and utilization records for the MWD team. These records include *DA Forms 2807-R* and *3992-R*. Once an additional month has passed, the latest month will be transferred to the MWD team folder in the filing system.

TRANSPORTATION REQUIREMENTS

2-50. Every unit, installation, and activity with MWD teams should assign enough support vehicles so that all missions and kennel support activities can be properly conducted and maintained. The minimum standard for transportation is one support vehicle per two MWD teams (two dogs and two handlers). Vehicle requirements should be determined by mission requirements and the overall number of MWD teams. Some modifications are necessary so that the MWD can be seen and can provide a visual deterrence during law and order or force protection and antiterrorism missions. These modifications will maximize the mission effectiveness of the dog while providing the dog with a comfortable, supportive, and protective area inside the vehicle. This type of modification is normally a kennel crate or cage secured in the rear of the vehicle; the crate or cage should have a nonslippery surface.

2-51. MWD handlers perform complex missions. Four-by-four type passenger vehicles should be used to ensure that MWD teams can get to any mission during inclement weather or on rough terrain.

2-52. Pickup trucks are essential for the transportation of explosive training aids. *AR 190-12* lists the requirements for explosives transported in government vehicles. However, pickup trucks should not be used for transporting MWDs for training or during road patrol type duties. This vehicle offers limited protection for the MWD from inclement weather and innocent bystanders.

2-53. Warning signs stating "CAUTION—MILITARY WORKING DOGS" should be placed on all sides and the rear of any and all vehicles used to transport MWDs. The signs should also be printed in the host nation language. The only exception to this is the use of unmarked vehicles for President of the United States and Vice-President of the United States missions.

NARCOTICS AND EXPLOSIVE TRAINING AIDS

2-54. Detector dog training and evaluations require the frequent use of narcotic and explosive training aids. Due to the potential safety hazards and risks involved with these aids, it is absolutely mandatory for all MWD personnel and commanders to know and follow the regulatory guidance. MWD leaders and handlers must participate in periodic review and training on the requirements noted in *AR 190-12*. These requirements are clearly stated in *AR 190-12*.

Chapter 3

Managing Military Working Dog Operations

Managing an MWD program requires attention from the program manager, the KM, and the plans NCO. This chapter will help these individuals better understand their specific requirements and concerns regarding the scheduling of various missions.

SCHEDULING AND EMPLOYMENT

3-1. The KM or plans NCO assists and advises the commander or operations officer in scheduling and employing MWD teams. The KM and plans NCO are the most qualified to determine the duty schedules, work cycles, and hours of operation to support each mission. All MWD teams should be utilized at least 24 hours per week to perform nontraining related duties.

3-2. Scheduling must allow for a minimum of four hours of patrol proficiency training per week and four hours of detector training per week. Details concerning proficiency training requirements are discussed in *Chapter 2*. Daily scheduling should include enough time for feeding, grooming, and exercising the MWDs as well as maintaining equipment and kennel facilities.

PREDEPLOYMENT OPERATIONS

3-3. MWD leaders and planners need to ensure that handlers are prepared for all possible missions before the moment of execution. Careful planning and concurrent training is an ongoing process that involves all soldiers in every kennel section. No MWD team should ever be sent on a mission without the proper preparation.

PROGRAM MANAGERS PREDEPLOYMENT RESPONSIBILITIES

3-4. MACOM program managers play a key role in MWD team readiness. The careful management of information concerning each kennel enables leaders and planners to keep the unit in a constant state of readiness for any possible deployment or no-notice mission.

3-5. Program managers keep track of the—

- Total number of personnel at each kennel in the MACOM.
- Certification status of all teams at each kennel in the MACOM.
- Deployability status of all MWDs and handlers (see *Chapter 8*).
- Total number and location of deployed MWD teams.
- Status of kennel facilities at deployment locations.
- Status of VCO support at deployment locations.

KENNEL MASTER PREDEPLOYMENT RESPONSIBILITIES

3-6. The KM's responsibility is to ensure the daily accomplishment of all missions. The readiness of all assigned MWD teams is a process that requires continuous effort from all leaders. KMs must maintain statistics and make the necessary corrections to have all personnel and equipment available for future deployments and missions.

3-7. The KM is responsible for—

- Managing the number of personnel assigned to each kennel.
- Verifying the status of certified teams and the dates of certification.
- Maintaining the deployability status of MWDs and handlers.
- Ensuring that soldiers have soldier readiness packets prepared according to *AR 600-8-101*.
- Conducting monthly and quarterly counseling.
- Maintaining accurate and detailed training records that will give the forward deployed KM an understanding of strengths and weaknesses.
- Ensuring that MWD handler equipment for short-notice deployments is prepared at all times.
- Conducting annual 9-millimeter handgun qualification. More than one qualification should be conducted (if possible) to better train MWD handlers on the use of the 9-millimeter handgun and to ensure that handlers are always qualified.
- Ensuring that handlers meet the standards outlined in *AR 600-9* and *FM 21-20* by conducting height and weight testing and physical fitness tests.
- Ensuring that the mandatory common military training and common task tests are conducted according to *AR 350-1*.
- Training soldiers in tasks associated with deployments (for example, kennel setup, utilization, and missions).
- Ensuring that MWD handlers have received any needed driver's licenses or transportation motor pool (TMP) licenses before a deployment order arrives.
- Ensuring that soldiers have received the needed individual readiness training in preparation for deployment in a combat zone.
- Conducting mine awareness training.

Plans Noncommissioned Officer Predeployment Responsibilities

3-8. Deployment planning must include coordination with the supporting military VCO. The plans NCO must consider the expected duration of the deployment and the types of facilities anticipated at the destination.

3-9. The plans NCO is responsible for ensuring that the handlers and MWDs in the MWD team or kennel section are technically proficient. The plans NCO is best able to evaluate and recommend to the KM which MWD team is most qualified and prepared for deployments and missions. Through continuous and realistic predeployment training, the plans NCO ensures that the MWD teams are proficient in—

- Crowd control training.
- Building searches.
- Tactical movements (through obstacles and from point to point).
- Gunfire training (handler and decoy).
- Defense training.
- Controlled aggression.
- Explosive detection (area searches, bunker searches, and tactical vehicle searches).
- Narcotics detection (area searches, bunker searches, and tactical vehicle searches).

Additional Predeployment Considerations

3-10. Army mobility requirements necessitate that each MWD has a shipping crate. Shipping crates are used to transport MWDs, simplify handling and loading, and give MWDs adequate space and ventilation during transport. Shipping crates may also be used as temporary kennels at intermediate sites or at the deployment destination.

3-11. The average water requirement for a single MWD is 10 gallons a day. Enough dog food must be taken to last 90 days or until resupply can be established. MWDs must be fed the standard high-performance dog food, contracted and supplied by the General Services Administration (GSA), unless otherwise directed by the supporting VCO. MWDs are fed twice daily unless the supporting VCO directs another meal frequency.

3-12. First aid kits must accompany the MWD team on every mission. Information on the deployment destination, climate, and terrain help determine the specific contents of the first aid kits. For example, if MWDs are employed on rough terrain, a pad toughener is included in the first aid kit.

3-13. All mission-related equipment must travel and arrive with MWD teams so that they can be fully operational as quickly as possible. Any equipment that is not sent with the MWD teams should be sent by other means to arrive at the location within 60 days of the MWD team's arrival. An example equipment list is in *Appendix E*.

DEPLOYMENT CONSIDERATIONS

3-14. There are many important factors to consider when leaders and MWD planners are preparing for deployments. To ensure that every deployment results in mission success, every aspect from notification to redeployment must be considered. These aspects include, but are not limited to, field kennels, training, and operations.

Field Kennels and Support Facilities

3-15. A suitable kennel site must be carefully selected. Before deployment, some possible locations may be selected on a map from a leader's reconnaissance. As soon as possible after arrival, the sites should be inspected to choose the most suitable location. When selecting a site, consult a VCO (whenever possible) about possible health hazards.

3-16. Locating the kennel in a congested area may be necessary to protect it from enemy attack. A temporary screen or physical barrier may be needed to block the dog's sight so he will be able to rest.

3-17. When shipping crates are used as temporary kennels, the interior, exterior, and ground beneath the crate must be cleaned daily to prevent the accumulation of moisture and waste or insect infestation. It must be raised four to six inches off the ground to allow for adequate drainage and to reduce parasite breeding places. In hot climates, place crates under trees, a tarpaulin, or plywood to provide shade and ventilation. Spread gravel (if available) around and under the crates to allow for drainage and the easy removal of solid waste. A large tent or prefabricated building can serve as a temporary storage location for equipment, rations, and other supplies.

3-18. A second tent or building erected at the kennel site to house handlers and the KM provides the kennel with sufficient manpower for security of the kennel area. Although conditions around a temporary kennel may be primitive, a high standard of sanitation is essential to prevent diseases and parasite infections. Permanent kennel facilities must be erected within one year of the arrival of MWD teams to the location of operational use. Modular expandable MWD transport containers (NSN 961120, 961130, or an equivalent) (*Figure 3-1*, page 3-4) with integrated air conditioning, heating, and equipment storage are recommended for use as temporary kennels in field environments.

Deployment Training

3-19. As soon as possible after arriving at the deployment destination and while making sure that site security is established, deployment training must begin. Deployment training allows the MWD teams to quickly develop familiarity with their new location. This is also known as right-seat ride time. The KM or plans NCO needs to evaluate or reevaluate the degree of skill of each of the MWD teams so that the KM and NCO are fully prepared to support the deployed force. Initial evaluation and training should consist primarily of field problems such as attack, search, re-attack, and scouting and basic obedience. Training should also include the handler firing his assigned weapon while maintaining control of the MWD. Some training should be done at dusk, and then in darkness, so that the handler may see (for safety) and the MWD may adjust to the noise and flash of weapons.

Figure 3-1. Modular, Expandable MWD Transport Container

MANEUVER AND MOBILITY SUPPORT OPERATIONS

3-20. MMSO consist of measures necessary to enhance combat movement and the ability to conduct movement of friendly resources in all environments. These measures ensure that commanders receive personnel, equipment, and supplies as needed. MMSO is conducted across the full spectrum of military operations. The primary focus of military police during MMSO is to ensure the swift and uninterrupted movement of combat power and logistical support. KM teams will coordinate MWD operations with the plans and operations officer or task force commander.

AREA SECURITY OPERATIONS

3-21. When setting up listening posts or conducting combat patrols, consider the availability and suitability of PDs for the particular mission. The KM or plans NCO must be briefed early regarding the mission. This will ensure that the most suitable PD team can be selected and that the handler can prepare for the mission. The handler should check the MWD for any limiting ailments and conduct training rehearsals to become familiar with the mission. The KM or plans NCO and selected handlers must be involved in planning conferences and briefings involving the utilization of an MWD team.

3-22. In defensive operations, combat patrols are used to provide early warning, to confirm intelligence information, and to detect or deter enemy action. PD teams greatly enhance the security of reconnaissance and combat patrols.

3-23. On combat patrols, a PD works at maximum efficiency for only two or three hours. During inclement or extreme hot or cold weather, this time will vary. The KM, plans NCO, and VCO should advise the commander on the MWDs' particular capabilities. The team is most effective in uninhabited areas. If an MWD frequently alerts on friendly forces and is continuously taken off his alert, he soon loses interest and reliability. Only the best teams should be selected.

Defiles and Route Reconnaissance/Surveillance Operations

3-24. MWD teams can be used to enhance the security at defile and route reconnaissance/surveillance operations. Consider the following to ensure mission success:

- The mission at hand (the possible threat scenario).
- The size of the area to be covered.

- The type and number of MWD teams.
- The duration of the mission.
- Sustainment supplies.
- Communications.

3-25. Operational planning must ensure that the proper numbers and types of teams are requested to perform the mission. Planning considerations must allow for proper rotations and rest-to-work ratios for the MWD teams.

Checkpoints

3-26. MWD teams are often called on to enhance the effectiveness of checkpoints. After receiving the mission to perform checkpoint duty, planners and handlers need to consider the following:

- The force protection condition (FPCON) (see *Appendix F* for details).
- The type and number of teams required versus those available.
- Sustainment equipment.
- Communications.
- Available support elements.

Area Defense

3-27. Areas that cannot be covered by static defensive posts because of vegetation, terrain, or some other peculiar feature can be secured using PD teams. Prior to positioning any MWD team in a defensive posture, ensure that intelligence preparation of the battlefield (IPB) details any possible placement of friendly or enemy mines. When relieving others from established defensive positions, use range cards to familiarize the team with the area.

3-28. A PD team can detect personnel using concealed avenues of approach and provide an early warning to defenders. Each MWD team's area of responsibility should be large enough so that the team can move to take advantage of the prevailing wind directions. A fixed position should be established to provide covering fire for each MWD team in case the team must withdraw closer to the defensive perimeter after the MWD has alerted on an attempted hostile probe or penetration.

WARNING AND RESPONSE PROCEDURES

3-29. Several procedures should be developed for warning friendly forces of a PD's response because circumstances may prevent the use of a radio. Also, different types of responses tailored to fit a variety of potential situations should be developed. For example, it may be desirable to have the MWD team follow the alert and locate the cause. At other times, combat intelligence may require the team to maintain the alert position until assistance arrives or to withdraw to a more advantageous position. Whatever action is taken, the handler should not release an MWD unless it is necessary to defend—

- Himself.
- Other personnel.
- Protected resources.

INTERNMENT AND RESETTLEMENT OPERATIONS

3-30. I/R operations consist of those measures necessary to guard, protect, and account for people that are captured, detained, confined, or evacuated by US forces. In any military operation involving US forces, accountability and the safe and humane treatment of detainees is essential. US policy demands that all persons who are captured, interned, evacuated, or held by US forces are treated humanely. This policy applies from the moment detainees become the responsibility of US forces and continues until the time they are released or repatriated. (See *ARs 190-8* and *190-47* and *FMs 3-19.40* and *27-10*.)

3-31. MWDs offer a psychological and actual deterrent against physical threats presented by personnel housed in an I/R facility. They cannot be used as security measures against US military prisoners. MWDs reinforce security measures against penetration and attack by small enemy forces that may be operating in the area. MWDs also provide a positive, effective alternative to using firearms when preventing disturbances or escapes by personnel.

3-32. MWD handlers will not use their MWDs to guard prisoners inside prisons or detainee holding facilities and confinement facilities. In addition, MWD handlers will not use their MWDs to degrade, torture, injure, or mistreat EPWs, detained personnel, civilian internees (CIs), or other detainees in US custody.

Note: MWDs are not used during interrogation operations.

Demonstrations

3-33. Conduct periodic demonstrations in full view of personnel housed in an I/R facility to increase the psychological deterrent of an MWD. Emphasize how easily and quickly an MWD can overtake a fleeing individual. Highlight the MWD's ability to attack and overcome physical resistance. Demonstrate an MWD's scouting ability. To ensure a successful demonstration, use only the most qualified MWD teams.

Perimeter Security

3-34. Use an MWD team as an addition to perimeter security by making periodic, unscheduled patrols around the perimeter fence during periods of darkness. During inclement weather, a temporary blackout or an electrical failure increases the number and frequency of patrols.

WARNING

Never set patterns or establish routines when conducting security patrols. Enemy observing security measures can exploit set times by attacking when the patrol ends or before it begins.

Narcotic and/or Explosives Detection

3-35. Walk an MWD team through living areas to search for contraband. PEDD teams can be utilized to detect explosive devices and components in the I/R facility and PNDD teams can be utilized to detect narcotics.

Work Details

3-36. Position an MWD team between the work detail and the area offering the greatest avenue of escape. MWD teams provide a valuable addition to work detail guards, particularly those employed in areas offering the greatest potential for escape. However, MWD teams are deterrents and must not be used as guards.

Police Intelligence Operations

3-37. *AR 525-13* directs all military law enforcement to develop foreign and domestic criminal intelligence. MWD teams are an integral part of the police intelligence operations (PIO) process. PIO are conducted in conjunction with all functions in which the MWD teams perform. For example, a PEDD team working at a checkpoint to detect explosives will also be gathering intelligence. This holds true for all MWD functions.

3-38. Leaders and planners of MWD operations must develop procedures to ensure that the MWD teams gather and submit all intelligence in a timely manner. Standardized reports are an effective means that allow for the accurate and detailed submission of intelligence. Contact the supporting intelligence section (S2), provost marshal office (PMO), or CID office to ensure that all PIO coordination is in place prior to performing MWD functions.

LAW AND ORDER OPERATIONS

3-39. Law and order operations are those that MWD teams perform to enforce laws, orders, and regulations and to maintain discipline and safety. Law and order operations can be performed in both combat and peacetime environments

RANDOM ANTITERRORISM MEASURES PROGRAM

3-40. The random antiterrorism measures program (RAMP) is designed to be used at all FPCON levels to increase physical security awareness throughout an installation. RAMP involves the random use of measures, identified in *AR 525-13* and *TC 19-210*, for FPCON levels to make installation operations less predictable and therefore more secure.

3-41. The use of MWD teams during RAMP provides for more thorough detection of narcotics and/or explosives and for a psychological deterrent. An MWD presence can also increase confidence in installation security in general.

3-42. KMs coordinate with the PMO or S3 to ensure that RAMP checks are successful. Coordination will include—

- The reconnaissance of patrol areas (map reconnaissance as a minimum).
- Rehearsals and inspections before operations.
- Transportation.
- Equipment needed on site.
- A threat briefing.
- Rules of engagement.
- Use-of-force criteria.
- After-action review (AAR) requirements (see *Appendix D*).

TEMPORARY DUTY MISSIONS (UNITED STATES CUSTOMS SERVICE/HIGH-RISK PERSONNEL SECURITY MISSIONS)

3-43. USCS safeguards the internal security of the United States through border searches. Because of this, they have broad powers in conducting searches. By utilizing an MWD team, all vehicles, ships, luggage, and baggage can be checked at border clearance crossings or ports. MWD teams are called upon to support local or federal authorities in detection operations. The Joint Enforcement Military Customs Operation places MWDs at border clearance ports to conduct searches of vehicles and cargo entering the United States. The MWD team takes no law enforcement action. The MWD team only searches and reports. *AR 90-12* gives a more detailed explanation of civil support by MWD teams.

3-44. The Air Force has been designated to serve as the primary service point of contact for all very important person (VIP) missions received from the Office of the Secretary of Defense, USSS, and/or Department of State. The VIP coordination officer who (under authority from the Secretary of Defense) tasks and coordinates all missions performed by DOD PEDD team personnel, regardless of service affiliation, issues all mission tasks.

3-45. MWD leaders and planners ensure that the certification of the team selected to perform the mission is well documented and current. Other considerations in planning for USCS support include—

- The travel arrangements of the MWD team.
- Government travel credit cards.
- Lodging for the handler.

- The kennel and boarding.
- VCO support.

Military Working Dog Public Demonstrations

3-46. Public demonstrations are conducted to show the abilities of MWD teams. These demonstrations are coordinated through the KM to ensure that the appropriate team is available for the target audience. The goal of public demonstrations is to deter criminal activity and to promote public acceptance of the MWD.

3-47. At no time should demonstrations make an MWD a form of entertainment. The KM or plans NCO should ensure that the details of sensitive police training techniques are not explained to the public. Avoid lengthy or frequent public demonstrations that allow MWDs to tire easily or become bored.

Health and Welfare Demonstrations

3-48. Initially, the MWD team should provide the responsible commander with a basic demonstration of the MWD's ability. The KM or plans NCO places a training aid in an undisclosed location, and the handler enters the search area with the MWD. The handler presents and conducts a systematic search of the area. When the MWD alerts, the appropriate reward is given. This demonstrates the MWD team's proficiency.

3-49. After the demonstration has been conducted, the unit should ensure that the area being searched is free of soldiers and that all rooms are unlocked. MWD teams are a tool the commander uses to find any illicit narcotic or explosive devices. It is the responsibility of the unit to properly search the room after the MWD has cleared the room. If the MWD team responds, military police personnel are called to continue the search. The MWD handler conducting the search should not search for the illicit narcotics or explosives. Unit commanders should also ensure that no personnel enter the building or area until the search is complete. This is usually accomplished by placing NCOs at various entranceways to stop personnel from entering the area.

Competitive Events

3-50. MWD teams should be encouraged and allowed to participate in competitive events, seminars, or conferences conducted by outside agencies or police canine organizations. Exposure to competitive events can challenge MWD teams to develop skills or learn advanced employment and training techniques. These events may require new tasks or methods that the handler can use to identify and apply to military missions. In this way, the team's proficiency and effectiveness are enhanced.

Recovery Operations

3-51. After every deployment or extended off-site mission, MWD leaders need to ensure that vital actions are conducted upon the return of the MWD team. These actions include—

- Conducting an AAR after every MWD deployment or extended mission to assist in capturing lessons learned. See *Appendix G* for detailed information on AARs.
- Conducting veterinary inspections.
- Returning any explosive or narcotic training aids.
- Updating training requirements and records.
- Recertifying teams.

Chapter 4

Legal Considerations

The use of MWDs has the same legal considerations as the use of a military police soldier. Each situation and location may be subject to various Status of Forces Agreements (SOFAs), general orders, and other legal restrictions. The following chapter addresses some fundamental areas. It is important to consult with the Staff Judge Advocate whenever legal questions pertaining to an MWD arise.

USE OF FORCE

4-1. Procedures for the use of force are found in *AR 190-14*. This regulation states that personnel engaged in law enforcement or security duties will use the minimum amount of force necessary to control the situation. This regulation further details the levels of force and encourages commanders to substitute nonlethal measures for firearms when they are considered adequate for safely performing force protection, antiterrorism, and law and order functions. The regulation states that in evaluating the degree of force required for specific situations, the following options should be considered in the order listed:

- Verbal persuasion.
- Unarmed defense techniques.
- Chemical aerosol irritant projectors (subject to host nation or local restrictions).
- The military police baton.
- MWDs.
- The presentation of a deadly force capability.
- Deadly force.

4-2. Although high in the order of measures, it is clear that the release of the MWD as a use of force is a nonlethal option to be used before deadly force is considered. It must be understood that the release of an MWD to pursue, bite, and hold a person is likely to result in an injury to the suspect. This injury can be very minor such as scratches or bruises; however, serious bodily harm is possible. For this reason, it is required by *AR 190-14* that all lesser means of force must be used first whenever possible.

4-3. The use of force differs between MWD handlers and regular military police because of the uniqueness of the MWD. Positive control of the MWD must be maintained at all times throughout all of the use-of-force measures. Often, just the presence or arrival of a handler with his MWD can control a situation. If the desired level of control is not achieved, verbal persuasion is the first degree of force to be used. If verbal persuasion is not enough force to maintain or gain control, the next measure may be considered. If the situation allows, military police may use unarmed defense techniques, spray irritants, and the baton before MWD handlers use the MWD as a measure of force.

4-4. A handler is not likely to use unarmed defense techniques when holding an MWD on leash. The handler must remain in positive control of the MWD; therefore he will not be able to get close enough to a suspect to employ unarmed defense techniques. The use of chemical aerosol irritant projectors or the baton would also hinder the handler's ability to keep positive control of the MWD. Therefore, the options are not reasonable measures of force available to the handler.

4-5. Military police and task force commanders, along with PMs, establish clear policies regarding the release of MWDs according to *ARs 190-12* and *190-14*. If the minimum amount of force necessary to control the situation requires the use of MWD force off leash, the handler must ensure that the following steps are followed:

- Ensure that the MWD has identified the same target as the handler before releasing him to bite.
- Give the warning order, "Halt, halt, halt, or I will release my dog." If assigned in a foreign country, give the order in the primary language of the host nation.
- Warn all bystanders to cease all movement.
- Call the MWD off the pursuit and regain control of him if the suspect indicates surrender or stops moving.

4-6. Use extreme caution when removing an MWD from a suspect. Regain and maintain leash control of the MWD until he has become calm enough to obey the commands "Heel" and "Stay." Do not release an MWD if the suspect is out of sight, unless an off-leash building search is authorized. Do not release MWDs in areas where children are present.

4-7. MWD teams are not used for crowd control or direct confrontation with demonstrators unless the responsible commander determines it to be absolutely necessary. When it is necessary, dogs are kept on a short leash to minimize the danger to innocent persons. Dogs are not released into a crowd. Civil disturbance contingency plans include specific criteria for the use of MWD teams that are consistent with *FM 3-19.15, AR 190-12,* and *AR 190–14*.

SEARCHES

4-8. Rules 311 and 313 through 316, *Manual for Courts-Martial (MCM),* prescribe HQDA policies concerning the conduct of searches and inspections and the disposition of property seized in conjunction with those activities. Property refers to property of the United States or of nonappropriated fund activities of an armed force of the United States wherever it is located.

4-9. Commanders and law enforcement personnel must be aware of the distinction between searches, seizures, inspections, and inventories and the provisions for the execution of these actions as outlined in the *MCM*. Knowledge of this information is necessary to ensure the legal use of a PNDD or PEDD.

4-10. Probable cause is a reasonable belief that a crime has been committed and that the person, property, or evidence sought in connection with the crime is located in the place or on the person to be searched. Probable cause must be established prior to searching any persons, property, or residence. Probable cause can be established in the following ways:

- Personal observation.
- Information obtained from a reliable informant or from the witness who observed the crime.
- Use of a detector dog team.
- Any combination of these three.

4-11. The detector dog team is considered a "reliable informant," according to Rule 315 (f)(2), *MCM*. The reliability of a team is established through the initial and annual certifications of the detector dog team by the certification authority and through proper documentation of the MWD's training records.

4-12. Certification establishes the foundation for an officer to grant a search authorization or warrant. Detector dogs are first certified after finishing their basic training. To meet the legal requirements permitting their operational use, each detector dog team must undergo an annual certification and a validation.

4-13. Each detector dog team undergoes a certification when first assigned to duty. A designated certification authority conducts the certification. The detector dog team must perform a demonstration for the person who authorizes any search. This may be any commander from company level up to the post commander. The guideline is that the person who witnesses the demonstration can only authorize a search of units he has command authority over. This person must also review the probable cause folder of the dog and must validate the team as reliable, in writing. This validation is good as long as the commander is assigned to that command.

4-14. The certification authority should witness a demonstration of the dog's capabilities. All substances the dog has been taught to respond on should be used. Additionally, the dog's capability to detect past/present odor should be demonstrated. At the completion of the demonstration, the results are documented in letter form, signed by the commander, and entered into the dog's probable cause folder. The MWD handler is responsible for preparing and maintaining the probable cause folder and the contents (see *paragraph 2-49*, page 2-9). The probable cause folder accompanies the handler each time the dog is used on an actual search. Each document in the folder should be explained to the commander requesting a detector dog search. The commander with command authority should witness a simple demonstration of the dog's capabilities prior to a search. The above procedure should be followed whenever possible. This procedure establishes the dog's reliability on a continuing basis and is essential in establishing probable cause.

4-15. PEDD and PNDD teams must maintain high standards of accuracy to provide the desired level of confidence in their searches. PNDD teams must maintain a 90 percent accuracy rate for the number of trials conducted, while PEDD teams must maintain a 95 percent accuracy rate for the number of trials conducted. The installation commander authorizes the support of local and federal law enforcement agencies. *DOD Instruction 3025.12* governs the use of all PEDD teams. It governs the use of PNDD teams and ensures compliance with *Section 1385, Title 18, United States Code (USC) (The Posse Comitatus Act) (18 USC 1385 [The Posse Comitatus Act])*. Complete certification, recertification, and decertification details for narcotic and explosive detector MWD teams can be found in *AR 190-12*.

Chapter 5
Patrol Dogs and Characteristics

PDs can be used in traditional law and order functions, but commanders and planners should not overlook the tactical enhancements that a PD team offers to combat support operations. All aspects of military police functions can be enhanced by the employment of MWD teams. This chapter discusses many practical applications of the PD team.

EMPLOYMENT TECHNIQUES DURING COMBAT SUPPORT OPERATIONS

5-1. PD teams can perform many combat support operations. These teams enable the tactical commander to free up soldiers and employ their resources in other areas. This force multiplier is especially valuable in area security and force protection and antiterrorism operations.

PERIMETER SECURITY

5-2. PD teams are useful for perimeter (*Figure 5-1,* page 5-2) and distant support posts because they are usually located away from normal activity and large numbers of people. These posts also usually enclose large areas, and it would take a large number of single sentries to secure the area effectively. The large size of these areas allows the PD team to change positions to take advantage of the prevailing wind direction. These posts may be secured only during periods of advanced security and high threat. Occasional random posting of these areas is recommended, especially during periods of limited visibility (rain, snow, dust, smoke, and fog) or night conditions. Barriers and obstacles (such as fences, buildings, gullies, and streams) must also be identified and considered in security post planning. Place the PD team so that these obstacles offer the least interference to security.

5-3. Perimeter security, depending on the area to be covered, should use a two-team concept. The teams overlap each other while walking around the perimeter. PD teams should not come into direct contact with each other. Considerations include—

- The commander's intent.
- Priority intelligence requirements.
- Rules of engagement.
- Use-of-force criteria.
- The wind direction and distracters in the area (see *Chapter 8*).
- Friendly and enemy forces around the area.
- Rest areas within the perimeter to allow the MWD to relieve himself.
- Watchtowers or other security forces in the area.

5-4. The PD team should initiate the perimeter security operation by walking in a zigzag pattern around the exterior perimeter. The team should be allowed to go outside of the security area to ensure that no unwanted intruders are present.

Figure 5-1. Perimeter Security

INTRUSION DETECTION SYSTEM AUGMENTATION

5-5. Continuing development and use of Intrusion Detection Systems (IDSs) for security make it necessary to reassess the role of PD teams in the security of priority resources where IDSs are used. Although IDSs are designed to greatly reduce the probability of a successful area penetration, they are not designed to replace PD teams. IDSs cannot see beyond the outer clear zones, counter intrusions, or scout the progress of an intruder force after penetration. They are also subject to malfunction and breakdown. PD teams cannot counter all potential problems, but they can be used selectively to counter some of the IDS limitations. With the advice of the KM, security planners should develop plans that set an effective balance between sensor systems and PD teams. Some suggested functions for integrating PD teams into IDS-augmented security systems are as follows:

- PD teams may be used inside or outside critical sites or locations of HRP type personnel areas to give immediate response to areas not clearly seen by closed circuit television (CCTV) or to support security response teams in their role of alarm assessment
- PD teams may be used to scout hidden intruders from the point of penetration to the hiding location inside protected areas.
- PD teams may be posted outside critical sites during periods of increased threat to expand the protected area beyond the range of the IDS. This use is particularly desirable when there is heavy vegetation beyond the perimeter clear zone or when protected areas are located on or near installation perimeters and may, therefore, be easily accessible to the public.
- PD teams may be used as an element of the security response team or as an element of other designated response forces.
- PD teams may be used to compensate for decreased IDS effectiveness when poor visibility or other environmental conditions adversely affect the IDS equipment or when the IDS is not functioning.

Secondary Response Forces

5-6. Secondary response forces are typically used as blocking forces and for sweeps and counterattacks. PD teams can greatly improve the capabilities of the secondary response force. Some factors that must be considered are that—

- Blocking positions must be selected downwind from the PD team's position when PDs are used with blocking forces. In this type of mission, PDs should only be used to alert the forces of possible intruders.
- The secondary response force moves into the wind with the PD team in the lead position when the PDs are used for sweeps. It is imperative that security be placed on the PD teams to ensure that they are protected.
- A PD team can often pinpoint the exact location of individual intruders.

5-7. A PD team can help detect, alert, and catch hidden intruders while clearing a position that has been counterattacked successfully. PEDDs can be used to help clear the area of explosives. (See *Chapter 7*.)

Listening Posts

5-8. On a listening post, the PD team should be positioned forward of the tactical area to reduce distractions to the PD, yet close enough to maintain contact with friendly forces. In selecting a location, consider the primary detection senses of the PD. Whenever possible, PD team listening posts should be located downwind from any potential enemy position or dismount avenue of approach. Other locations force the PD to rely only on his hearing and sight for detection, reducing his effective use of scent.

Combat Patrols

5-9. In defensive operations, combat patrols are used to provide early warning, to confirm intelligence information, and to detect or deter enemy action. PD teams greatly enhance the security of reconnaissance and combat patrols.

5-10. On combat patrols, a PD works at maximum efficiency for only two or three hours. The team is most effective in uninhabited areas. If a PD frequently alerts on friendly forces and is continuously taken off his alert, he soon loses interest and reliability.

5-11. Patrol leaders and members must be briefed on actions to take when a dog handler is seriously wounded or killed. Dogs that have worked closely with patrol members and have developed a tolerance for one or more of them will usually allow one of the patrol members to return it to the kennel area. However, the dog may refuse to allow anyone near his handler, and other handlers may need to be called to help. The canine must be separated from the handler by—

- Coaxing the dog away with friendly words or food.
- Covering the dog with a poncho to immobilize it.
- Staking out the dog or leashing him.

5-12. A human will always take precedence over an animal. If no effort is effective, the dog may be destroyed. However, all efforts should be made to avoid this; an MWD is an expensive and valuable asset to replace.

5-13. If a situation arises where a canine is injured but the handler is not, the handler must be allowed to accompany the canine. If wounded or killed, an MWD should be evacuated using the same assets and should receive the same consideration as that given to a soldier under the same circumstances.

Tactical Patrols

5-14. PD teams can be used in tactical patrols to detect enemy presence, to help the patrol avoid discovery, and to locate enemy outposts. When the PD alerts, the handler should signal the patrol to halt until the cause of the alert can be identified and the patrol can proceed safely. If a firefight develops while the PD team is at the point position, the PD team should respond to fire team's directions and act as a regular member of the patrol.

5-15. Generally, the best locations for PD teams are directly in front of the patrol or on the flanks to enhance the flexibility and security of the formation. Prevailing wind direction should be used to improve the patrol's chances for early warning of enemy forces. If the wind direction is from the rear of the patrol, the PD is forced to rely on sight and sound and may not be as effective. The handler must concentrate on the PD to read his alert and will not be able to use a weapon rapidly. Therefore, members of the patrol should be assigned to protect the PD team.

5-16. PD teams can be used to enhance the capabilities of an ambush patrol. A PD team should be placed in front of the patrol to minimize distractions, yet close enough to maintain contact with other patrol members. PDs on ambush patrols must remain silent and not respond aggressively to approaching enemy forces.

INTERNMENT AND RESETTLEMENT OPERATIONS

5-17. PD teams are used during I/R operations for camp perimeter security and to guard against escapes. The PD's keen senses of sight, smell, and hearing assist the handler and camp authorities in detecting unauthorized activity. A PD team's presence also acts as a deterrent to escapes. In the event of an escape, the PD is a valuable asset that can assist in the search and recovery of internees.

PD Duties and an Internment and Resettlement Camp in Iraq

On 13 April 2003, SGT Carey A. Ford and his MWD Rex responded to a riot at Camp Bucca, Iraq. It was believed that an Iraqi internee had taken a pistol belt and some personal equipment from a guard. The prisoner had taken this equipment back to his tent and the 600 other internees refused to cooperate with the guard force. SGT Ford and MWD Rex went into the I/R camp unarmed. The MWD team moved through the disruptive and unruly crowd providing a secure environment for the quick-reaction force to extract the prisoner and retrieve the stolen pistol belt and personal equipment. SGT Ford related that EPWs would not come anywhere near his MWD Rex. "We work that fear to our advantage."

Four days later, SGT Ford and Rex were called on to scout for and apprehend an escaped EPW. The fugitive had a substantial lead on them, having made his escape two hours prior to SGT Ford's notification. SGT Ford entered a possible unexploded ordnance and/or antipersonnel minefield with Rex, who was also trained to detect explosives. The MWD team was determined to locate the dangerous fugitive and prevent access to any possible weapons caches since many had been located in the area. After hours of relentless searching, SGT Ford and Rex located and apprehended the escapee. Their actions provided the psychological deterrent necessary to prevent future escape attempts.

RIOT/CROWD CONTROL

5-18. Extreme caution should be applied before committing PDs to a potential crowd control situation. Although the presence of a PD can be a powerful statement to an enemy or terrorist, it can elicit a reverse and potentially adverse effect on individuals in a riot/crowd control situation.

5-19. It is difficult to employ PDs properly during demonstrations or riots due to the high levels of confusion and excitement and the large number of potential antagonists. PDs should be trained to remain undistracted by loud noises and crowds. A crowd's reaction to a PD in a riot environment may be dependent on the location. In some areas of the world, the appearance of PDs at an angry crowd scene results in an escalation of violence. The crowd will often challenge soldiers to use the PDs as a measure of force. This is particularly true if a situation can be manufactured or provoked so as to be interpreted as an unreasonable use of force by the military. The presence of PDs can cause extreme fear (deterrence) in some cultures and regions. This fear can be advantageous and should be exploited when needed.

5-20. Committed PD teams may be used as an alternative to the use of deadly force to gain control of a riot situation. If the commander located on the ground directs the employment of PD teams for direct confrontation with demonstrators, PD teams should be held in reserve, out of the sight of the crowd. As the situation escalates, PD teams may be moved forward to within the sight of the crowd, but well away from the front lines.

5-21. When committed to the front lines in direct confrontation, PDs are allowed to bite only under the specific circumstances authorized by the responsible commander. PDs should never be released freely into a crowd. Other riot control force personnel should be positioned approximately three meters from any PD team when conducting any form of riot control. This is to prevent unintentional bite injuries to friendly, riot control force personnel that may happen to get between a PD and an instigator participating in the riot or demonstration.

5-22. PD teams must move a safe distance from the crowd to ensure the safety of the PD. Use of PD teams for direct confrontation with demonstrators is not recommended. According to the Army's use-of-force policy, the application of physical force against a person who cannot reasonably be suspected of committing an offense is not appropriate. Presence in a crowd does not necessarily constitute an offense or a situation where it would be reasonable to apply the degree of physical force of a PD. A PD would not be used unless the person is engaged in some other readily apparent criminal behavior such as theft, destruction of property, or assault.

EMPLOYMENT TECHNIQUES DURING FORCE PROTECTION AND ANTITERRORISM OPERATIONS

5-23. PD teams contribute significantly to the military police force protection and antiterrorism effort. PDs are directly responsible for executing specific security-related measures and for providing support to other government agencies that contribute to force protection.

5-24. MWD support is accomplished under the leadership of the respective command PM. Force protection is accomplished through an active role in physical security, HRP security, law enforcement, and antiterrorism efforts. Force protection and antiterrorism operations consist of those actions that prevent or mitigate hostile actions against DOD personnel (to include family members), resources, facilities, and critical information. During force protection operations, offensive and defensive measures are coordinated and synchronized to enable the joint force to perform while degrading opportunities for the enemy.

5-25. In operations, PD teams are used to enhance the detection capabilities of the combat support force and to provide a psychological deterrent to hostile intrusions. If the PD teams are properly located in a hostile environment, the PD's alert provides an initial warning to the presence of a hostile force. In past combat operations, PD teams often provided warning of attacks early enough to allow a response force time to deploy and to prevent enemy forces from reaching their objectives. PD teams have also helped clear protected areas of hostile persons, explosives, and weapons after attacks.

Crowd Control in Kosovo

During Task Force Falcon (2A) in Kosovo, LTC Cloy (the battalion commander of 1-36 INF, 1 Bde, and 1AD) requested MWD support to assist the Alpha company clear a main supply route (MSR) that had been blocked by 10,000 Kosvar Albanians. LTC Cloy had already deployed both quick-reaction forces to assist Company A because the crowd would not stay off the MSR. LTC Cloy called the battalion operations center and directed that the MWD teams be sent to the scene.

SFC Barnes, SSG Rogers, SSG Barters, and SGT McClintock responded with their dogs Queeny, Danny, Sandor, and Fedor and linked up with an infantry squad. The MWD teams and infantry squad moved dismounted to the town of Gjilani. The crowd was starting to become violent. LTC Cloy was in the middle of the crowd, standing in the center of the city at the critical MSR intersection. LTC Cloy attached the MWD teams to Company A. CPT Curtain, the company commander, told SFC Barnes to move the crowd away from his area and to establish a perimeter. The MWD teams moved into the diamond formation with the dogs on guard. Splitting the crowd in half, the MWD teams moved through the crowd and reached the city center MSR intersection in the center of the city. The handlers then used the dogs to move the crowd back out of the road. Company A and the four MWD teams established a perimeter line to keep the MSR open and traffic moving. They also apprehended troublemakers in the crowd.

Later that evening when the dog handlers entered into the DFAC, LTC Cloy stood up, gained everyone's attention, and stated, "SFC Barnes, you and your handlers were like Moses. Your dogs parted the Red Sea!"

ACCESS CONTROL POINTS

5-26. A PD team may be used to enhance force protection and antiterrorism efforts at ACPs. The PD team's function at an ACP is to deter unauthorized access, crime, and terrorism. The handler that is assigned to the ACP should use the PD so that the public can observe and hear the dog. Care must be taken to ensure that the handler constantly keeps the PD under control to prevent incident or injury to innocent personnel.

BUILDING AND AREA SEARCHES

5-27. A PD team is especially effective in checking and searching buildings such as commissaries, post exchanges, finance offices, banks, and warehouses. The PD team can check buildings visually while patrolling and can stop and dismount so that the handler can physically check doors and windows. These checks should be made with the PD on leash. To take maximum advantage of the PD's scenting ability, the handler should approach buildings from downwind.

5-28. If a building is open or forced entry is evident, PD teams may be used to track hidden intruders from the point of penetration to their location. PDs may be released to apprehend an intruder suspected of committing a serious offense when the only alternatives are escape or the use of deadly force.

5-29. Prior to searching a building, the handler will determine whether to make the search with the PD on or off leash. Some of the factors that must be considered include the following:

- Friendly forces or innocent bystanders in the area.
- The time of day or night.
- Evidence of forced entry.
- The type and size of the building or area to be searched.
- The danger to any innocent persons in the area.
- The PD's ability to work off leash.
- The threat level and IPB.

5-30. Prior to releasing the PD inside a building or enclosed area, the handler will give a clear voice warning to any persons inside the building or in the area. They will be told to come out immediately and that failure to comply with these instructions will result in the release of a trained MWD. An interpreter may be needed to assist the PD team in giving the warning in the host nation language. All persons will be warned that the PD may attack without warning and that they could receive physical injuries. The handler maintains voice control of his PD throughout the search.

> ***Note***: A warning similar to the one for off-leash searches will precede on-leash searches. The searching team should be followed by at least one other military policeman to protect the search team.

APPREHENSION OF SUBJECTS

5-31. The use of PDs to attack and hold an individual enemy or terrorist should not be overlooked. A PD can be sent into a building, confined space, or cave to find, attack, and hold an individual until the handler and friendly forces can arrive to place the subject into custody. Although disadvantages do exist (injury or death of the PD), the PD is better-suited than a human to immediately find and subdue a subject in an area where the interior structure is unknown to friendly forces.

WALKING PATROLS

5-32. The PD team's contribution to MWD team law enforcement efforts is most effective when the team is on foot. Some of the law enforcement duties that a PD team can do as a walking patrol include checking or clearing buildings and patrolling parking lots, dependent housing areas, and troop billet areas.

5-33. Since it is easily seen, a PD used during daylight in congested areas provides a good psychological deterrent to certain crimes. PDs are tolerant of people, and the presence of a large number of people does not significantly reduce the PD's effectiveness except when conducting detection missions.

5-34. Using the PD team during darkness or periods of reduced visibility and in areas where there are fewer distractions can enhance the PD's detection ability. Military police may not detect a person fleeing a crime scene at night, but a PD will normally be able to detect and locate a fleeing suspect. When necessary and appropriate, the PD is also much more effective at pursuing and holding the suspect.

5-35. Periodic use of PD teams around on–post schools, especially when school is starting and dismissing, may help deter potential vandalism, child molestation, exhibition, and illegal drug activities. PD teams can also provide effective security for communications facilities, military equipment, and command posts (CPs).

ALARM RESPONSES

5-36. In responding to an alarm condition at facilities such as clubs, finance offices, or banks, the PD team should be among the first military police on the scene. The PD team may be used to search and clear the building and immediate area and help with any apprehensions. If the PD must be used to find a suspect, other persons should attempt to avoid contaminating the area with scents or tracks that may confuse him.

VEHICLE PARKING LOTS

5-37. PD teams are effective in detecting and apprehending thieves and vandals in vehicle parking lots. The presence of the PD team may deter potential acts of theft and vandalism. The PD team can be most effective by alternating between vehicle and foot patrols of the parking lots. During hours of darkness, when there is no activity in a parking lot, the PD team should approach the lot from the downwind side. If the PD alerts, the handler should locate and challenge the suspect for identification.

5-38. If a suspect tries to escape or avoid apprehension and the handler is reasonably certain the suspect has committed or is attempting to commit a serious offense, the handler should give a clear warning to the suspect to halt. When lesser means of force have failed to stop the fleeing suspect, the PD may be released to pursue, attack, and hold. If it is not safe to release the PD, the handler may attempt to follow the suspect if the PD has the ability to track.

Chapter 6

Narcotics Detector Dogs and Characteristics

MWDs are trained to perform dual roles. In addition to the uses and characteristics of the PD, the PNDD is trained to detect the odor and presence of specific drugs. This specialized ability to detect certain illegal drugs makes the PNDD a valuable tool to help commanders and other government agencies maintain law and order. The PNDD can be employed in combat and/or force protection and antiterrorism operations worldwide.

EMPLOYMENT FOR COMBAT SUPPORT OPERATIONS

6-1. The PNDD provides commanders with unique capabilities in combat support environments. The narcotic detection capabilities of these dogs can be used to maintain the order and discipline of US soldiers as well as combatants and other persons involved in operations (EPWs, CIs, refugees, or other detained/interred persons [during time of war/conflict]). In addition to narcotic detection functions, PNDD teams should not be overlooked for their PD functions as discussed in *Chapter 5*.

INTERNMENT AND RESETTLEMENT OPERATIONS

6-2. In addition to perimeter and other security duties, PNDD teams assigned to I/R camps can be used to ensure that no illegal drugs are at the I/R camps by conducting searches of all areas. Commanders and planners should consider having all inbound supplies or equipment searched by PNDD teams. Inbound and outbound mail and vehicles should also be considered as a means for the movement of illegal drugs and, as such, subject to a search by PNDD teams.

POSTAL OPERATIONS

6-3. Postal inspections, including mail storage areas and drop boxes, can be conducted in conjunction with a military customs inspector (MCI) and according to *DOD Directive 5030.49*. The postmaster or MCI can request PNDD support from the area PM or military police brigade S3 with coordination through the KM team.

6-4. A PNDD team can be used to search mail that is received or shipped. Searches can be conducted as the mail comes off or before it goes on the truck prior to being shipped. The PNDD team can search up and down the conveyer belt as the mail is removed from or placed on the truck. If a conveyer belt is not being used, the handler can instruct postal personnel how to arrange the mail in a secure area so that the PNDD team can search it more effectively. When a PNDD team is used in the close quarters of a mailroom or delivery area, all nonessential personnel should remain clear of the area while the team is actively working.

6-5. When the PNDD responds to a parcel indicating the possible presence of illegal drugs, the CID or the military police investigations section will be notified for further action. At this point the PNDD team's actions are finished.

AIRCRAFT AND LUGGAGE SEARCHES

6-6. Before beginning the search, ensure that all personnel have exited the aircraft and all compartments are opened. The PNDD team searches the aircraft before searching any luggage or cargo. When approaching the aircraft, conduct a fresh-air perimeter search from the downwind side. Once a perimeter search has been conducted, the team boards the vessel at either the front or rear. Once the area has been cleared to search, the handler ensures that only one door is open to limit draft and wind current. A systematic search of all areas within the aircraft is conducted from the top to the bottom. The handler advises other military police on the scene where the PNDD displayed a positive response, and the augmenting military police conduct a detailed search of the area surrounding the response.

6-7. When conducting inbound baggage inspections, the items should be aligned in a sterile area in rows with at least six feet between rows. The handler conducts an on-leash search, presenting each piece of baggage. When searching cargo, every effort should be made to search the cargo before it is palletized. The cargo should be aligned in rows, with at lest six feet between rows, in a sterile area (same as baggage). The handler conducts an on-leash search, presenting cargo as required. Once the PNDD team has searched the cargo, it may be palletized while remaining in the sterile area prior to embarkation.

6-8. Conducting debarkation operations is the same as inbound operations. The host unit (to which the cargo belongs or is being delivered) provides a representative (E6 or above) and a work detail to assist the PNDD team. When pallets are removed from the aircraft, they are placed in a sterile area. Once in the sterile area, the work detail unloads the pallets and arranges the cargo, as required by the PNDD handler.

FORCE PROTECTION AND ANTITERRORISM OPERATIONS

6-9. PNDD teams are extremely useful in force protection and antiterrorism operations. Commanders and other government agencies use PNDD teams to protect the community from the harmful effects of illegal drugs. PNDD teams protect the force and surrounding communities by deterring criminal activity and enhancing law enforcement capabilities.

HEALTH AND WELFARE INSPECTIONS

6-10. PNDD teams conduct health and welfare inspections when requested by the commander and coordinated through the KM, PMO, and S3. The commander contacts the KM directly to schedule a health and welfare inspection. The KM, in conjunction with the installation PM, the CID special agent in charge, and the military police investigations supervisor, briefs all levels of the command on the availability and capabilities of the PNDD teams available to them on the installation.

6-11. When scheduling the inspection, the commander should have the following information available:

- The requested time of inspection.
- The areas to be inspected.
- The number of living quarters.
- The number and size of common areas.
- Other pertinent information.

Appendix H provides a guide to assist handlers with health and welfare inspections.

6-12. Once the coordinating instructions have been established, the KM coordinates a time and location to brief the commander and demonstrate the PNDD team's capabilities. It is important that the unit requesting the health and welfare inspection practices operational security (OPSEC) and informs only those personnel who have an immediate need to know. This usually consists of the commander and first sergeant or sergeant major. The KM advises the PM and the military police investigation supervisor when the inspection is scheduled and if any additional support will be needed.

6-13. A systematic inspection of the living quarters and common areas is conducted. If the PNDD responds to a locked wall locker, the soldier assigned to that wall locker is summoned to open it. Once the wall locker has been opened, unit representatives conduct a detailed search of the wall locker and its contents.

DEPARTMENT OF DEFENSE SCHOOLS AND DRUG ABUSE RESISTANCE EDUCATION PROGRAM INSPECTIONS

6-14. KMs may receive requests from Drug Abuse Resistance Education (DARE) officers or DOD school principles to conduct a health and welfare inspection of a school. When conducting school inspections in conjunction with DARE, basic building inspection techniques apply. The primary concern when inspecting schools is the safety of the students and the secrecy of the search.

6-15. When possible, searches should be conducted in conjunction with a school activity, such as an assembly or fire drill. This will allow the handlers to conduct the search with a decreased risk of student involvement. Another option is to search the school after normal school hours or on a weekend or holiday. In any event, the use of the PNDD team on school grounds must be cleared by the school administration and the PM must be notified in advance.

ASSISTANCE TO MILITARY POLICE INVESTIGATIONS SECTION AND/OR THE DRUG SUPPRESSION TEAM

6-16. The military police investigation section or the drug suppression team (DST) may contact the KM to schedule an inspection for a number of reasons. PNDD teams may be requested to conduct an inspection of a residence or building where illegal drugs have already been found. In addition, unit commanders may request a PNDD team to inspect the quarters of a soldier who has shown a positive response on a urinalysis. The PNDD is not used to verify or test a substance that has already been found. PNDD handlers contact the military police investigations section or the DST in any circumstance that a PNDD has responded positively according to local standard operating procedures (SOPs).

MILITARY CUSTOMS MISSIONS

6-17. The US Army has been designated as the proponent service for the military customs program. According to current regulations, PNDD teams are used to inspect aircraft cargo areas and airframes when the military customs inspector program manager (MCIPM) directs these inspections. When conducting military customs missions, the handler should first inspect the vessel, then if deemed necessary by an MCI, continue with baggage.

6-18. When scheduling military customs missions, the MCIPM coordinates with the KM according to local SOPs. At a minimum, the MCIPM notifies the KM of—

- The requested time and date of the search.
- The number and types of vessels to be searched.
- The amount of cargo to be searched.
- The number of passengers and crew requiring the MCI's clearance.
- The determination on whether the search is a high-risk or selective search.

6-19. PNDD teams may inspect vessels such as military or civilian aircraft, boats or ships, vehicles, rail cars, or any other vehicle capable of carrying passengers and/or cargo according to the military customs program. These inspections are primarily conducted outside the continental United States. However, some missions in the continental US may be necessary for units returning from foreign theaters where the installation is their first stop within the customs territory of the United States. In addition, the military customs program may require preclearance for units or individuals deploying to areas outside the customs territory of the United States where an MCI has not yet been assigned or is not available. When conducting preclearance of mobilized or deploying soldiers' personal baggage while in formation, the MCI must ensure that the soldiers do not distract the PNDD team.

VESSELS, BAGGAGE, AND CARGO INSPECTIONS

6-20. When establishing sterile areas, the MCI coordinates with the PNDD handler and ensures that the area is of sufficient size with minimal distractions to ensure the optimal working environment for the PNDD. When scheduling working parties for embarkation or debarkation, the PNDD handler coordinates with the MCI to ensure that sufficient personnel will be present to accomplish the mission in a timely manner, but not so many soldiers that they might present a distraction to the PNDD. If an examination has been deemed necessary by the MCI, the handler presents each piece of baggage or equipment to the PNDD.

UNITED STATES CUSTOMS SERVICE MISSIONS

6-21. Missions in support of the USCS come from the MWD program manager via the KM. Handlers must have—

- Active government credit cards.
- Current passports.
- Appropriate civilian attire (or a civilian clothing allowance).
- Complete transportation requirements.

6-22. When temporarily assigned to the USCS, local USCS agents, policies, and procedures direct the PNDD team. Weapons are not authorized for handlers assigned to the USCS. The uniform of the day is determined by the USCS and is normally civilian attire.

6-23. Prior to deploying, every attempt should be made to conduct pretraining exercises with the USCS. These pretraining exercises will allow the MWD to acclimatize to larger quantities of contraband, which may be found while on the mission. The USCS will provide large training aids and determine the PNDD saturation point prior to deploying.

6-24. PNDD teams can expect to perform searches on aircraft, baggage, and tractor trailers as well as vehicle lines at US borders. When assigned to the USCS, the handler does not conduct physical searches of any kind. If the PNDD responds to an area, the handler advises a USCS agent and continues with the search, leaving the agent to conduct the actual physical search of the area.

6-25. Tractor trailers are searched in a USCS sterile area. Once the cargo has been offloaded and placed in standard lines, the PNDD team performs an exterior inspection of the vehicle before moving to all interior compartments and concluding with a systematic search of the cargo.

6-26. While assigned to a US border, the PNDD may be employed in vehicle lines. Vehicle lines are generally 10 to 20 vehicles in a row (placed bumper to bumper) that are inside a sterile area. The PNDD team travels up one side of the vehicle line and back down the other. The handler presents all productive areas. In the event that a USCS agent selects a vehicle for an individual vehicle inspection, that vehicle will be placed in a sterile area and the driver will open all the compartments of that vehicle. Once all occupants are clear of the vehicle, the PNDD will conduct a systematic inspection of the entire vehicle.

6-27. The PNDD will only be used in his narcotics detector role while deployed in support of the USCS or other federal agency. The PD capabilities of the PNDD will not be used for any reason. If the handler is instructed to use his PNDD in any method other than that of a narcotics detector dog, he must immediately notify his KM.

PNDD Support to US Customs Counterdrug Missions

In the summer of 2003, SGTs Payne and Mason from Fort Knox, KY, and SPC Johnson from Fort Sill, OK, were tasked to support a US Customs counterdrug mission along the southwest border between Mexico and the United States. After an overview and train up, the PNDD handlers and their assigned dogs went to work with the customs agents to detect illegal drugs at various points of entry into the country.

These teams became very proficient in detecting large quantities of illegal drugs. The PNDD teams detected narcotics in false floors, dashboards, gas tanks, seats, and hidden panels of many vehicles.

After three months of searching vehicles at checkpoints, the PNDD teams were responsible for detecting over 4,800 pounds of marijuana, over 95 pounds of cocaine, and other undisclosed contraband.

RANDOM VEHICLE SEARCHES

6-28. The use of a PNDD at ACPs and installation gates can increase installation antiterrorism capabilities by providing a visual deterrent to terrorist groups and other criminal elements. The installation commander, task force commander, or base camp commander may task the PM to conduct random vehicle inspections according to local SOPs and the FPCON. When this occurs, the PM submits a schedule of random vehicle inspections to be conducted within a time frame that includes times, dates, and locations of inspections to be conducted. The schedule will be approved or disapproved and the inspection parameters will be dictated (will every vehicle be inspected or will vehicles be inspected at random according to the guidance provided) by the commander and returned to the PM. The PM then notifies the KM accordingly and provides him with a copy of the schedule with the commander's signature and inspection parameters. The PNDD's patrol capabilities may be used at ACPs to search for stowaways and personnel trying to gain access to the installation or a controlled area by hiding in delivery vehicles or other large cargo type vehicles. Extreme caution must be used when using the team in this manner. The commander must be aware that once the PNDD is given the command to search for personnel the MWDs aggression level will increase, thus increasing the chances of the PNDD biting personnel.

ON-CALL DRUG RESPONSE

6-29. PNDD handlers will normally be placed in an on-call status. On call means the handler must remain within response time and distance according to the local SOP. When on call, PNDD handlers do not consume alcoholic beverages; however, they are in a duty position that allows them to wear appropriate civilian attire as well as leave the post according to the local SOPs.

6-30. When on call, PNDD handlers maintain positive control of a secondary communication device issued by the command. The secondary communication device may be a digital pager or cellular telephone as determined by the command. When recalled, the PNDD handler returns to the kennel in the most expedient legal manner to retrieve his PNDD. The PNDD will only be transported in an approved government owned vehicle (GOV). At no time will an MWD be transported in a POV. The handler may respond in appropriate civilian attire according to the local SOP.

6-31. KMs provide insight and guidance to the PM on technical matters of the SOP. The KM should provide recommendations as to acceptable distances which may be traveled while on call, required response times, the required secondary mode of communication, and any installation-specific items of interest, such as an alternate uniform based upon the customary civilian attire that is prominent in the area.

Chapter 7

Explosive Detector Dogs and Characteristics

In addition to the uses and characteristics of the PD, the PEDD is also trained to detect the odor and presence of explosives. This specialized ability to detect explosives makes the PEDD a valuable asset to help commanders and other government agencies maintain law and order, while enhancing safety procedures. PEDD teams can be employed in combat and/or force protection and antiterrorism operations worldwide. PEDD teams are valuable assets. The task force commander, the PM, and military police planners should consider the use of PEDD teams during combat support operations. This chapter details sound employment techniques and the capabilities of the PEDD teams that enable leaders to develop plans.

INTERNMENT AND RESETTLEMENT OPERATIONS

7-1. The Army is the DOD executive agent for all EPW/CI operations. The military police is tasked with coordinating shelter, protection, accountability, and sustainment for I/R operations. The I/R function addresses military police roles with US military prisoners, EPWs, CIs, refugees, or other detained persons (during time of war/conflict). Although the combat support military police unit initially handles EPWs/CIs, modular military police (I/R) battalions with assigned military police guard companies and supporting MWD teams are equipped and trained to perform I/R operations for sustained amounts of time (see *FM 3-19.40*).

7-2. PEDD teams may be used for searching internees' belongings or confinement cells. PEDD teams should not be used to search a person. A PEDD is trained to attack with or without command; any sudden movements the internee may make could be perceived by the PEDD as an aggressive act. Safety must remain a priority for both the handler and military personnel.

7-3. The commander can also use PEDD assets to search the property of attached military personnel when needed. This is especially helpful when new units are rotating or replacing old units. These types of searches ensure that no explosives are taken out or brought in by US forces.

POSTAL OPERATIONS

7-4. Postal inspections, including mail storage areas and drop boxes, can be conducted in conjunction with an MCI and according to *DOD Directive 5030.49*. The postmaster or MCI can request PEDD support from the area PM or military police brigade S3 with coordination through the KM team.

7-5. The PEDD team can be used to search mail that is received or shipped. Searches can be conducted as the mail comes off or before it goes on the truck prior to being shipped. The PEDD can search up and down the conveyer belt as the mail is removed from or placed on the truck. If a conveyer belt is not being used, the handler can instruct postal personnel how to arrange the mail in a secure area so that the dog can search the mail more effectively. When a PEDD team is used in the close quarters of a mailroom or delivery area, all nonessential personnel should remain clear of the area while the team is actively working.

7-6. When the dog responds to a parcel, indicating the possible presence of explosives, the area is evacuated immediately. Military police secure the area and set up a temporary CP a safe distance from the suspected explosives. Use the stand-off distance according to *Appendix I*, as applicable. Use the PEDD team to search the safe area for any possible secondary explosives. Explosive ordnance disposal (EOD) personnel are requested to dispose of the explosives or render them safe. EOD personnel may give explosives and improvised explosive device (IED) components to federal agencies for further investigation or turn over the scene to federal agencies after explosive hazards have been neutralized. The PEDD team remains at the temporary CP and resumes detection operations once EOD personnel remove the explosives.

CHECKPOINTS AND ROADBLOCKS

7-7. PEDD teams can be used at checkpoints and roadblocks to supplement the control and movement of vehicles, personnel, and materiel (*Figure 7-1*). PEDD teams are added to security forces to detect explosives and to prevent illegal actions that may aid the enemy. Using these control measures deters terrorist activities, saboteurs, and other threats.

Figure 7-1. Checkpoint Search

7-8. When using PEDD teams at hasty or deliberate checkpoints and roadblocks, a search area must be designated for vehicles and a separate area must be designated for the vehicle occupants or any dismounted personnel. The PEDD team conducts a search of the vehicle only after security forces at the checkpoint ensure that all of the occupants have exited the vehicle and that a threat to friendly forces does not exist. Security forces at the checkpoint provide overwatch while the PEDD team searches for explosives.

Note: At no time should a PEDD handler be used to search the interior of a vehicle. PEDD handlers are trained on searching techniques with a PEDD and should not be used for security force tasks.

7-9. In the event that the dog indicates the presence of explosives, the PEDD team withdraws immediately from the vehicle. The area is evacuated and EOD personnel and the chain of command are notified. The PEDD team stays behind a barrier at the safe-distance location in case the EOD team leader requests further detector assistance. (See *Appendix I* for stand-off distances.)

Note: Be sure that radio operations are stopped until the PEDD team and security forces reach a safe distance from the suspected explosives.

CORDON AND SEARCH OPERATIONS

7-10. Military police leaders and other task force commanders should consider using PEDD teams when conducting cordon and search operations. PEDDs are trained and able to greatly improve the combat force's success in uncovering hidden explosives and people.

7-11. Once the military police or other security forces have established the cordon, the PEDD teams should move in and operate with the search elements. During the search, the PEDD teams provide warning of hidden explosives or persons while the rest of the search element employs overwatch and search techniques.

7-12. The search element leader directs the actions of the team if people or explosives are detected. The PEDD handler should be prepared to advise the commander on the capabilities of the dog and actions required by the rest of the team to ensure the PEDD team's safety. It is important that all members of cordon and search teams rehearse together before conducting operations.

FORCE PROTECTION AND ANTITERRORISM OPERATIONS

7-13. PEDD support for force protection and antiterrorism operations can be requested anywhere in the world and in different circumstances. In forward deployed areas or hostile environments, PEDD teams are often requested to ensure that any suspected explosives are detected before harm can come to US Army, joint, and allied forces. In peacetime environments around the world, PEDD teams perform the detection operations much the same as in combat support operations. The environment surrounding explosive detection operations does not interfere with the explosive detection capabilities of the PEDD team.

SUSPICIOUS/UNATTENDED PACKAGE RESPONSES

7-14. Once a suspicious/unattended item that may contain explosives has been discovered, the military police or security forces on the scene request a PEDD team. Normally, military police evacuate the area and establish a CP a safe distance away from the threat using the stand-off distance listed in *Appendix I* of this manual. The PEDD team checks the perimeter and secures the area for secondary explosives. EOD personnel should respond to the scene and make contact with the PEDD team to be briefed on the location of the package. The PEDD team moves to within 50 meters of the package and begins an area search up to package. The PEDD team also scans the immediate area for secondary devices, when applicable.

Note: At no time should a PEDD team be used to search the package itself. EOD personnel have received special training in handling suspicious/unattended packages and determining if a suspicious/unattended package contains explosives.

7-15. A PEDD handler must not present, touch, open, or physically disturb a package that is suspected of concealing explosives. Once the MWD team has discovered a suspicious item, EOD personnel assume responsibility for the area and take control of the situation in order to render the package safe or dispose of it as necessary.

PEDD Operations in Support of Operation Iraqi Freedom (OIF)

In April of 2003, SGT Raymond Rivera was performing explosives detection missions with his PEDD Kevin in support of OIF near Baghdad, Iraq. SGT Rivera was requested to sweep an area that a support element was in the process of clearing. There was so much debris that a forklift was being used to move large amounts of furniture and building supplies.

The PEDD team would search a portion of the area and then the forklift would come in and haul off the debris searched. After the third load was hauled off, the dog responded to a mattress on the ground with an office table on it. Expecting an explosive devise, SGT Rivera notified the NCOIC of the detail and the area was cleared of all personnel (to a safe distance) and secured. EOD personnel were notified of the suspected explosives and responded to the scene.

SGT Rivera was standing by at the safe area while EOD personnel uncovered two hand grenades that were buried in the sand just below the surface. SGT Rivera said, "Even though it was only two grenades, I was amazed that Kevin found them under all the debris. I'm glad we found them and not our soldiers."

BOMB THREATS

7-16. A PEDD team should be requested for assistance in detecting explosives whenever bomb threats are made against persons or property for which the Army is responsible. The PEDD team's search for explosives begins only after the area or building has been evacuated of all personnel and security has been established. EOD personnel only respond when a suspect item is found; they do not conduct or participate in searches except in support of the USSS. EOD units place a team on standby once the unit is alerted of a bomb threat. PEDD teams coordinate this with the local EOD unit prior to any operation.

7-17. The PEDD team, EOD personnel, and site leadership collaborate to resolve the situation at a nearby CP (see *Appendix I* for stand-off distances). When PEDD teams are located on an installation, their use must be included in the bomb threat planning guidance.

7-18. A systematic search by the PEDD team should include the assistance of a spotter to ensure safety and accuracy. PEDD teams are to be used only to detect the possible presence of explosives.

WARNING:

At no time will the handler touch, open, or otherwise disturb a possible explosive device. IEDs are easily disguised and can be triggered to explode by motion, sound, light, heat, or remote devices.

7-19. Handlers and spotters must not change or otherwise disturb anything in the search. If the lights or any electric device are off or on, do not touch the switch until a thorough search is conducted.

7-20. Ensure that the handler and spotter are provided with available safety equipment including—

- A flack vest (in addition to body armor).
- Flashlights or helmet-mounted lights.
- Gloves.
- Safety goggles.

> **WARNING:**
>
> **Always keep track of how much time has elapsed when using a PEDD team at bomb threat incidents. Sufficient time must be maintained to ensure that the team exits the building or area well before the announced or suspected detonation time.**

7-21. If the PEDD responds to the presence of explosives, the PEDD handler and spotter note the location and leave the area. EOD personnel then begin measures to render the device safe or remove it. On the request of EOD personnel, the PEDD team may then continue the search until another suspected explosive is found or until the area is cleared by EOD personnel.

7-22. If the search is 30 minutes or longer in duration, explosive training aids should be planted periodically so that the dog can find them. This helps to keep the dog interested in the search. Training aids should only be planted in areas that have already been searched by the PEDD team.

HIGH-RISK PERSONNEL SEARCHES

7-23. Military police provide the task force commander with valuable area security capabilities to protect C2 headquarters, equipment, and services essential for mission success. HRP security is an important task within the area security function. PEDD teams may be employed to enhance explosive detection and protection capabilities.

7-24. The KM or plans NCO determines which PEDD teams will perform the mission. The number of PEDD teams for any mission is determined by the threat, the number of PEDD teams available, any conflicting missions, and other environmental considerations noted in *Chapter 8*.

7-25. PEDD teams may be used as part of a combined arms operation or as an element of a larger security force. When used as part of a combined arms team, the PEDD teams need to be involved will all troop-leading procedures, including rehearsals. All personnel must have a complete understanding of the mission in order to create a safe and secure environment for the HRP. The PEDD teams provide explosive detection and personnel detection capabilities. PEDDs are also a deterrent to hostile forces or terrorists operating within the area of operations.

7-26. Prior to an HRP visit to an area, the PEDD team assists with force protection actions by searching for explosives and IEDs. PEDD also conduct searches of and provide security for outer perimeter areas. During HRP operations, the handler—

- Briefs the on-scene commander of the dog's capabilities.
- Communicates with EOD personnel.
- Obtains the HRP's itinerary and determines locations to be searched.
- Requests a spotter or helper to assist.
- Ensures that areas are secure after searches are done.
- Informs the commander of results and continues follow-on missions.

ACCESS CONTROL POINTS

7-27. Using PEDD teams at ACPs can increase an installation's or base camp's force protection and antiterrorism capabilities by providing a visual deterrent to terrorist groups and other criminal elements. The PM notifies the KM of the requirement to perform vehicle searches at ACPs.

7-28. In the event that the PEDD detects the presence of suspected explosives, the military police forces at the ACP secure and evacuate the area (see *Appendix I* for stand-off distances). EOD personnel are then notified to render the item safe or remove it. PEDD handlers do not attempt to remove or otherwise disturb any suspected explosives.

7-29. PEDD team patrol capabilities may be used at ACPs to search for stowaways and personnel trying to gain access to the installation or a controlled area by hiding in delivery vehicles or other large cargo type vehicles. Extreme caution must be used when utilizing the team in this manner. The commander must be aware that once the PEDD is given the command to search for personnel the potential for injury exists.

HEALTH AND WELFARE INSPECTIONS

7-30. PEDD teams conduct health and welfare inspections when requested by the commander and coordinated through the KM, PMO, and S3. The commander contacts the KM directly to schedule a health and welfare inspection. The KM, in conjunction with the installation PM, the CID special agent in charge, and the military police investigations supervisor, briefs all levels of the installation command on the availability and capabilities of the PEDD teams available to them.

7-31. When scheduling the inspection, the commander should have the following information available:

- The requested time of the inspection.
- The areas to be inspected.
- The number of living quarters.
- The number and size of common areas.
- Other pertinent information.

Appendix H provides a guide to assist handlers with health and welfare inspections.

7-32. Once the coordinating instructions have been established, the KM coordinates a time and location to brief the commander and demonstrate the PEDD's capabilities. It is important that the unit requesting the health and welfare inspection practice OPSEC and inform only those personnel who have an immediate need to know. This usually consists of the commander and 1SG or sergeant major. The KM advises the PM and the military police investigations supervisor when the inspection is scheduled and if any additional support is needed.

7-33. A systematic inspection of the living quarters and common areas is conducted. If the PEDD responds to a locked wall locker, the soldier assigned to that wall locker is summoned to open it. Once the wall locker has been opened, unit representatives conduct a detailed search of the wall locker and its contents.

AIRCRAFT AND LUGGAGE SEARCHES

7-34. Before beginning the search, ensure that all personnel have exited the aircraft and that all compartments are opened. The PEDD team should search the aircraft before searching any luggage or cargo. The PEDD team conducts a fresh-air perimeter search from the downwind side when approaching the aircraft. Once a perimeter search has been conducted, the team boards the aircraft at either the front or rear. Once the area has been deemed clear, the handler ensures that only one door is open to limit draft and wind current. A systematic search of all areas within the aircraft is conducted from the top to the bottom.

7-35. If the dog alerts the handler to a possible presence of explosives, military police or security forces evacuate and secure the scene. The PEDD team moves a safe distance away from the suspected explosives but remains available to further assist EOD personnel, if requested. Once EOD personnel clear the area, the PEDD team may resume the search until another suspected explosive is found or until the area is cleared by EOD personnel.

7-36. When conducting inbound baggage inspections, the items should be aligned in a sterile area in rows with at least six feet between them. The handler should conduct an on-leash search, presenting each piece of baggage. When searching cargo, every effort should be made to search the cargo before it is palletized. The cargo should be aligned in a sterile area in rows with at least six feet between them (the same as baggage). The handler should conduct an on-leash search, presenting cargo as required. Once the PEDD team has searched the cargo, it may be palletized while remaining in the sterile area prior to embarkation.

7-37. Conducting debarkation operations is the same as inbound operations. The host unit (to which the cargo belongs or is being delivered) provides a representative (E6 or above) and work detail to assist the PEDD team. When pallets are removed from the aircraft, they are placed in a sterile area. Once in the sterile area, the work detail unloads the pallets and arranges the cargo as required by the PEDD handler.

Chapter 8
Veterinary Support

At most locations, veterinary support is provided by an assigned Army VCO. When there is no assigned VCO, an attending VCO from a nearby military installation is assigned responsibility for providing veterinary support. The Army provides veterinary support for all MWDs as authorized by *DOD Directive 6400.4* and *ARs 40-1, 40-3*, and *40-905*. Emergency, civilian veterinary support is authorized for MWDs when there is not a VCO available and/or care is beyond the capability of the veterinary treatment facility (VTF). The responsible VCO and Army Veterinary Service chain of command are responsible for establishing standard procedures for approval, provision, and payment for emergency, civilian veterinary support.

VETERINARY RESPONSIBILITIES

8-1. The US Army Surgeon General provides professional veterinary support for the entire MWD program through the Army Veterinary Corps. This responsibility includes—

- Medical and surgical care at training facilities, bases, and installations.
- Inspections to ensure that the kennel facilities are sanitary.
- A professional review of plans for new construction and modification of kennels, support buildings, and sites.
- Prescribing an adequate feeding program.
- Instructing handlers and supervisors in all matters related to the health of MWDs.
- Conducting research to improve the MWD program.

8-2. The responsible VCO provides treatment for the diseased or injured MWD at the kennel site or at the installation VTF. The Army Veterinary Corps is responsible for equipping this facility and for providing medical and surgical supplies. The VCO is responsible for the MWD veterinary treatment record (VTR) and related information about examinations, immunizations, and treatment.

8-3. The VCO instructs handlers about dog health, care, feeding, and first aid. This instruction helps the handler to develop a better understanding of the MWD's health needs and improves the handler's ability to care for the MWD. The VCO prescribes an appropriate feeding program based on the MWD's health, the climate, and working conditions.

8-4. The VCO is involved in the review of all plans for new kennel construction or kennel modifications. This ensures that potential health and safety hazards can be corrected before construction begins.

VETERINARY INSPECTIONS

8-5. At least quarterly, the VCO inspects the MWD kennel facility to make sure that proper sanitary standards are being maintained. It is recommended that an animal care specialist conduct a monthly check of the kennels to ensure that no major safety violations are present. The VCO ensures that insect and rodent control efforts are adequate and that the handlers are properly maintaining the general health of the MWD. The quarterly inspection also includes the following elements:

- An examination of the kennel facilities for safety hazards and distractions that may interfere with the rest and relaxation of the MWDs.

- A review of the adequacy of the kennel structure (particularly for environmental conditions) and feeding and watering schedules. The VCO should make recommendations to help prevent disease and injury.
- A review of kennel sanitation and cleaning procedures and MWD food storage procedures to minimize the risk of disease transmission to the MWD.

MANDATORY TRAINING

8-6. The local VCO is required to provide training to the MWD handlers upon initial assignment and at least annually thereafter. More frequent training of handlers is recommended but not required. The Department of Defense Military Working Dog Veterinary Services (DODMWDVS) recommends that handler training be conducted at least quarterly. Topics include the following:

- Procedures on when and how to obtain veterinary support (an SOP is appropriate for this).
- General animal health.
- Grooming.
- Prevention of environmental injury (such as heatstroke or cold injuries).
- Emergency first aid (for bloat [gastric dilatation-volvulus], heat stroke, snake bites, and cold injuries).
- Administration or application of routine medications.
- Proper MWD feeding.
- Feed sanitation (proper storage, preparation, and cleanup).
- MWD chemical, biological, radiological, nuclear, and high-yield explosives protective measures.

8-7. It is best to schedule mandatory training so that seasonal problems are addressed prior to their occurrence and appropriate refresher training is provided prior to a deployment. This training is not meant to turn handlers into animal health technicians, but to give them the tools needed to provide emergency care until they can acquire veterinary support.

WEIGHT CHECKS

8-8. Each month the MWDs body weight must be recorded and sent to the responsible VCO. The KM may communicate this to the VTF. Body weight is also measured and recorded monthly in order to monitor body condition and detect important trends in weight loss or gain that may indicate illness. The KM and/or handler documents data such as food intake, stool consistency, and vomiting in the monthly records.

RECORDS/HEALTH CERTIFICATES

8-9. VTRs are maintained by the local VTF. It is very important that MWD handlers ensure that all health information is annotated in these records.

8-10. An examination is given just before shipping an MWD from one installation to another. The MWDs are examined to detect injury or disease and to support the safe shipment and continued good health and performance of the MWD. A health certificate is issued for the state or country into which the MWD is being shipped for TDY and must accompany the MWD. When the TDY period exceeds 10 days, a health certificate may be required for the MWD's return to his home station. A VCO signs this certificate no more than 7 days prior to reembarkation. If the TDY location is in a different Army VCO's area of operation, the gaining VCO is notified prior to the MWD's departure from the home station. An examination is given as soon as possible, but not more than 10 days after the arrival of an MWD at a new assignment, TDY, or deployed location and within 3 days of the MWD's return to his home station. It is important that the handler who deployed with the MWD is present for the examination so that the VCO can get accurate and complete information on any health issues while the MWD was deployed.

Feeding and Food Control

8-11. MWDs require a diet that is significantly different from that of pet dogs. Their work demands much higher levels of energy and larger quantities of essential nutrients. The standard high-performance diet, contracted and supplied by the GSA for MWDs, meets these nutritional needs and is the only approved feed unless otherwise directed by the attending veterinarian.

8-12. Special diets may be procured and fed to individual MWDs when the VCO determines that other than the standard diet is required. Normally, these diets are available from the VTF, but the MWD unit is responsible for procurement costs.

8-13. Some MWDs have been trained using a food reward schedule that also requires special food. Food reward is used only if an MWD fails to respond to any other type of reward. Procure the special food through normal supply channels or authorized local purchase. The amount of food each MWD should be fed depends on his weight, the amount of activity, and the climate. Local VCOs determine the proper amounts to feed. The responsible VCO must be told when an MWD is trained using food.

8-14. The VCO also prescribes the time of day each MWD is to be fed. This depends on the MWD's duty schedule and the schedule of other kennel activities. MWDs should be allowed 2 hours to eat; leftover food is disposed of within another two hours and feeding pans are cleaned and put away. Never leave uneaten food in the kennel past the authorized 2-4 hour feeding period. If the MWD finishes his meal prior to the end of the feeding period the feed pan may be removed.

Parasite Control

8-15. MWDs are at risk of becoming infested with internal parasites (intestinal worms and heartworms), and external parasites (fleas, ticks, and mange). Some of these parasites may also transmit infectious diseases to the MWD. In order to prevent infestation and disease transmission, all MWDs must be on a routine parasite prevention program supervised by the responsible veterinarian. At a minimum, this must include the administration of a monthly oral (pill or chewable) heartworm preventative and internal parasite preventive and the administration of a topical, veterinary-prescribed external parasite preventative against fleas and ticks.

8-16. When deployed to areas with high parasite risk, additional preventive measures such as prescription flea and tick collars and the oral antibiotic doxycycline will be prescribed by the supporting veterinarian. Doxycycline must be given with food. The handler must ensure that adequate supplies of all preventives are issued by the VTF prior to departing for TDY or deploying. Flea and tick collars will only be used when prescribed by the veterinarian and must be removed when the MWD is not under direct physical control (when off the choke chain and leash) of the handler to prevent accidental ingestion.

Food Storage

8-17. Food should be stored in a sealed container that will not allow contamination or rodents to enter the food (for example, pail, utility 7 gallon, NSN 7240-01-411-0581). Bulk storage of food must be in an area separate from the storage of other equipment, cleaning supplies, or materials that may contaminate the food. Bags should be elevated 4-6 inches above the ground, and neatly stacked with a 4-inch space between the stacks and the walls. The sanitation standards for the storage of MWD rations are identical to those required for the storage of bulk rations for personnel in a troop issue subsistence activity, exchange, or commissary.

Deployability Categories

8-18. The purpose of reporting MWD deployability status is to assist the MWD program managers, PM, commanders, and KMs with the selection of MWD teams for deployment and TDY and to rapidly identify MWDs with limited duty requirements. These guidelines take medical and physical fitness factors into consideration including—

- The anticipated MWD duty stress at deployed locations compared to the home location.
- The availability of veterinary support at the deployed location.

- The climate and environment of the deployed location.
- Any diagnosed or suspected medical conditions that may affect the MWD's performance.
- The MWD's physical conditioning and stamina as observed by the KM and handler.

8-19. It should be noted that some determinations are and will remain subjective; thus the need for prudent judgment by all individuals involved in these assessments remains. This is particularly true regarding the effect of age on an individual MWD. Old age is not a disease and MWDs are not downgraded just due to age; however, with increased age comes increased incidence of illness or disability. These problems may warrant a change in deployability status.

8-20. Deployability status is initially determined after the VCO examines the MWD at the receipt physical or semiannual physical exam. The KM determines training and proficiency status. The VCO and KM should discuss their findings and then categorize each MWD's status and forward the listing through the KM to the PM and MWD program manager. Updates to deployability status lists are performed at least monthly or as needed as changes occur in an individual MWD's status.

Category 1—Unrestricted Deployment (Both Outside the Continental United States and in the Continental United States)

8-21. The MWD is medically fit for any contingency or exercise. There are no limiting or compromising factors. Medical factors may exist but will not limit performance. Special diets, controlled drugs, or special medications can be sent with the MWD (or be carried by the handler) in a quantity sufficient to last the full duration of the anticipated deployment.

Category 2—Limited Deployment (Continental United States or Home Theater of Operation)

8-22. The MWD is medically fit for regions and missions with a minimal requirement for acclimation to heat or physical stress where complete veterinary support is available. The MWD is fit for short duration deployments. There are no significant limiting or compromising factors. Medical problems may exist which slightly limit performances but are controllable. Special diets or medications are required; however they are available in the deployed location or can be sent with the MWD. The reason for limited deployability must be reported.

Category 3—Temporarily Nondeployable

8-23. A medical condition exists that impedes daily duty performance and is under diagnosis, observation, or treatment. The MWD may still be allowed to work at the home station but in a limited capacity.

Category 4—Permanently Nondeployable

8-24. An unresolved medical or physical problem exists that frequently or regularly impedes daily duty performance and an early retirement date cannot be given. The reason for nondeployability must be reported. Medical or physical conditions warrant replacement within one year. The responsible veterinarian and KM should initiate an MWD disposition process at the time an MWD is declared category 4.

Kennel Sanitation Procedures

8-25. Cleanliness is one of the most important factors for the good health of an MWD. The KM must enforce sanitary measures in and around the kennel area. A good standard of sanitation is the result of the cooperation of the handlers, KMs, supervisors, and the VCO.

8-26. The VCO and the PM or military police commander should set the standard of sanitation. This standard must be maintained by each of the handlers and the KM.

8-27. Kennels must be sanitary, in a good state of repair, and thoroughly cleaned every day. Kennels should be disinfected at least once every week using only those disinfecting products approved by the VCO. Kennels should also be disinfected whenever an animal is removed from a kennel so that the kennel will be ready to be occupied by another animal.

8-28. Stool must be removed from the runs as often as necessary as but no less than twice daily to prevent the MWD from becoming dirty. The method of disposing of stool depends on local conditions and the type of sewage system present. If stool must be carried from the area in cans, the cans must be cleaned and disinfected after each use.

8-29. The entire kennel area must be free of refuse and garbage that could attract rats and insects. Mosquito control measures must be used in ditches and swampy areas in the vicinity of the kennels. Disinfectants and disinfectant procedures should be used only with the approval of the VCO.

MEDICATION

8-30. At times, a VCO prescribes special medication (to be given separately or mixed with food) for a sick or injured MWD. The handler must know how to administer these medications in both pill and liquid form.

8-31. When any foreign substance is placed directly into a dog's mouth, his first reflex is to spit it out. The handler must learn how to administer medication properly so that the MWD is forced to swallow. With the VCO's approval, pills or liquids can be mixed with a small amount of canned dog food or other soft food and fed to the MWD. To administer capsules or tablets—

- Place the fingers of your left hand over the MWD's muzzle and insert your left thumb under the lip and between his upper and lower teeth right behind the canine tooth, pressing your left thumb against the roof of his mouth to open it.
- Place the capsule or tablet into the MWD's throat at the extreme rear of the tongue to prevent him from spitting it out. Quickly remove your hand, close the MWD's mouth, and gently massage his throat until he swallows. The entire procedure must be quick and smooth to ease the MWD's apprehension and resentment.

8-32. Liquid medication is best administered with the help of another person. The handler—

- Holds the MWD's upper and lower jaws together with his left hand. The assistant pulls the MWD's lips away from the teeth at one corner of the mouth with his right hand. The MWD's nose is then pointed slightly upward, forming a natural funnel by the lip, and the assistant pours the liquid into this funnel.
- Elevates the MWD's head only slightly above the horizontal when giving liquid medicine. If the MWD's head is raised higher, he has difficulty in swallowing. Give the MWD adequate time for swallowing to prevent the liquid from getting into the trachea, nose, or lungs. Use extreme caution in giving oily liquids.
- Allows the MWD to lower his head and rest before proceeding if any signs of distress, such as coughing or struggling, appear. Do not give oral medications or any liquids if the MWD is unconscious or cannot swallow.

FIRST AID

8-33. Normally, the handler's early recognition of symptoms of illness or injury allows sufficient time to get assistance from a VCO. However, situations may arise when medical help is not immediately available. The seriousness of the incident may require that the handler take emergency actions to protect the life or health of the MWD. The first aid instructions that follow are commonly used to save life, prevent further injury, and reduce suffering from pain. In all emergency situations, notify the VCO as soon as possible so that the MWD can receive professional medical attention.

8-34. First aid kits should always be available in the kennel area and at training sites. The handler should also carry a first aid kit as part of his or her equipment for the MWD on all operational missions. The VCO will determine the contents of first aid kits. When items are used from the first aid kit, they should be replaced with new items immediately.

MUZZLES

8-35. When an MWD has been injured, the first consideration is to calm and immobilize the animal. Pain and distress, however, may cause the MWD to respond to the handler in an unpredictable manner. The MWD may not respond to verbal commands and may attempt to bite the handler and anyone helping the handler. Whether to apply a muzzle (*Figure 8-1*) or not depends on the nature of the emergency. If the MWD is unconscious, shows any difficulty in breathing, or has suspected head injuries, a muzzle should not be used. Otherwise, a muzzle should be used for safety.

Figure 8-1. Muzzle Use

8-36. There are several types of muzzles, but the leather basket muzzle is the best and most comfortable. It allows for free breathing and causes the least alarm and apprehension. The MWD can still inflict a wound while wearing this type of muzzle, so caution is required.

8-37. An improvised muzzle can be made using the MWD's leash and is called a leash muzzle. To apply the leash muzzle, tighten the choke chain on the MWD's neck by pulling the leash tightly with the right hand. Place the left hand, palm up, under the choke chain on the neck. Grasp the leash tightly as it passes through the palm of the left hand, wrap the leash once around the MWD's neck, and bring it up and across the left side of his head. Finally, wrap the leash twice around the muzzle and grab it tightly with the left hand.

CAUTION

The leash muzzle may be used when the leather muzzle is not available or when it would not provide adequate safety. Do not use the leash muzzle when the MWD is overheated, is having difficulty breathing, or is indicating that he may vomit. Do not leave it on for long periods in hot weather.

FRACTURES

8-38. If a fracture occurs, restrain the MWD and encourage it to lie down and remain quiet. If the face and jaw are not injured, put the MWD in a loose muzzle. Notify the veterinarian immediately when a fracture occurs or is suspected. If there is any bleeding, treat the MWD according to *paragraph 8-40,* then proceed with splinting and moving the MWD to the veterinarian if he is not immediately available.

8-39. Prior to moving the MWD, the fracture area needs to be immobilized with a splint. This should be possible for most leg fractures. The leg should be splinted in the same position it is placed by the MWD when it is hurt. Do not try to straighten or reduce the fracture. Splinting an MWD's leg is similar to splinting a soldier's leg and may be done with the same materials: bandages, torn cloth, splints from the medic's bag, or other available material including sticks, boards, and rolled newspaper. After the leg is splinted, or if the fracture cannot be splinted, gently move the MWD onto a litter or improvised litter, such as a piece of plywood, or into a kennel and carry it to a vehicle for transport. It is best to slide the litter under the MWD rather than lifting the MWD onto the stretcher. The MWD may need to be strapped to the litter to allow safe movement. Make sure to support the injured leg or area during movement.

Bleeding Wounds

8-40. Bleeding must be quickly controlled, particularly wounds in the foot or leg that are prone to bleed freely. Applying pressure directly on the wound may control bleeding; use a sterile bandage or clean handkerchief or pinch the edges of the wound with your fingers. As soon as possible, apply a pressure bandage.

Burns

8-41. Most burns occur when an animal comes into contact with hot water, grease, tar, or other scalding liquids. MWDs may also get electrical burns by chewing electrical wires. If an MWD is trapped in a burning building, it may suffer from smoke inhalation in addition to surface burns. The handler can prevent all of these situations through positive control of the MWD. If a burn does occur, applying cold-water soaks or ice packs to the burn for approximately 20 minutes may treat minor burns. The cold helps to reduce the pain. The hair around the burn should be clipped away and the burn should be washed gently with a surgical soap. Blot it dry with sterile or clean soft gauze or cloth. Protect the burned area from rubbing by applying antibiotic ointment and a loose fitting gauze dressing.

Shock

8-42. An animal may go into shock after injuries to internal organs, excessive bleeding, or trauma. Shock can be recognized by the following symptoms:

- A glassy look to the eyes.
- A rapid or weak pulse or rapid shallow breathing.
- A dropping body temperature. The lips and feet may feel cold.
- A paleness of the membranes of the mouth and eyes.
- Slow capillary refill time. To determine capillary refill time, press firmly against the MWD's gums until they turn white. Release the pressure and count the number of seconds until the gums return to their normal color. If it is more than two to three seconds, the MWD may be going into shock. Failure to return to the red-pink color at all indicates that the MWD may be in serious trouble and needs immediate assistance.

8-43. If a handler suspects that an MWD has internal injuries or is going into shock, request help from a VCO immediately. Keep the MWD warm and quiet, and lower his head to prevent possible brain damage. If it is necessary to move the MWD, use a litter. Administration of intravenous fluids by appropriately trained personnel will help to prevent permanent injury from shock.

Artificial Respiration

8-44. There are conditions that may cause an MWD to stop breathing. When this happens, do not panic. Follow these procedures:

- Open the MWD's mouth and check for any obstructions. Extend the MWD's tongue and examine his throat.
- Clear the MWD's mouth of any obstructions then, close his mouth, and hold it gently closed.
- Inhale, and then cover the MWD's nose and mouth with your mouth.
- Exhale gently. Do not blow hard. Carefully force air into the MWD's lungs and watch for his chest to expand. Repeat every five to six seconds or at a rate of 10 to 12 breathes per minute.

SNAKEBITES

8-45. The bite of a poisonous snake can cause serious illness and death if not treated immediately. Keep the MWD quiet and calm, and request veterinary assistance or move the MWD as quickly as possible to a treatment facility. Panic or exertion causes snake venom to move more rapidly through the bloodstream. If possible, kill the snake so it can be shown to the VCO. The handler should be careful not to be bitten by the snake or come into contact with the snake venom. Bites may occur on the face or neck of the MWD. When this happens, remove the choke chain and loosen or remove the collar and muzzle. Swelling occurs rapidly after a snakebite, and this equipment may restrict breathing. Position the MWD's head extended from his body to allow him to get maximum airflow. An ice pack applied to the bite area helps to slow the flow of blood and helps to keep the venom from spreading.

CAUTION

A tourniquet should not be applied. Constricting the blood flow could increase tissue death.

FOREIGN OBJECTS IN THE MOUTH

8-46. An MWD may get a stick or other foreign object stuck in his mouth or throat. The most common foreign object that becomes stuck is a Kong® or ball reward that is too small for the MWD. The MWD may cough, gag, have difficulty swallowing or breathing, paw at the mouth, and drool. If the MWD is having difficulty breathing, cautiously and gently open his mouth. Look for any abnormal object in the throat, under the tongue, between the teeth, in the gums, or in the roof of the mouth. If the MWD is having trouble breathing, removing the object is an emergency. The ball, Kong, or other object can be removed using pliers or clamps. If the object is not causing breathing problems, it is best to leave it in place and get veterinary assistance.

POISONOUS SUBSTANCES

8-47. There are many toxic chemicals an MWD may come in contact with and/or ingest. These include insecticides, herbicides, rodenticides, and antifreeze. The symptoms of poisoning vary. Unless the handler is certain the MWD has eaten a poison, do not treat for poison. Occasionally, MWDs accidentally swallow narcotics or explosives training aids. Exposure to poisons and accidental ingestion of training aides should be prevented by positive control of the MWD by the handler. If the handler knows that the MWD has eaten a poison, narcotic, or explosive, take the following actions:

- Request veterinary assistance immediately.
- Determine the type and quantity of poison, chemical, narcotic, or explosive that has been swallowed. If any part of the substance or container is available, keep it for the VCO to examine.
- If the Army VCO is not available, immediately make contact with a civilian veterinary to determine if vomiting is appropriate. Causing an animal to vomit is not recommended in all types of poisoning as it may cause more harm to the animal. When vomiting is appropriate, the animal may be induced to vomit by giving him one-quarter to one-half of a cup of hydrogen peroxide.
- Keep the MWD quiet and warm until the VCO arrives.

8-48. For future reference, keep a copy of the material safety data sheet for any training aide that may be ingested available. MWD first aid should be according to the material safety data sheet for people.

OVERHEATING

8-49. Overheating results when an MWD is unable to eliminate body heat rapidly enough. This condition requires immediate action by the handler to save the MWD's life. During hot, humid weather, an MWD may easily become overheated during training, during operations, or while being transported. A body temperature of 105° Fahrenheit or more, combined with poor response to commands, weakness, unsteady movement, vomiting, difficult or labored breathing, convulsions, and collapse are all symptoms of overheating. It is helpful to take the MWD's temperature in order to monitor the severity of overheating and progress in cooling, but temperature alone does not determine the need for or method of treatment.

8-50. For mild to moderate heat injury, the MWD's temperature is generally lower than 104° Fahrenheit. The MWD may have mild symptoms, to include uncontrolled panting, a rapid but somewhat weak pulse, mild to moderate difficulty breathing (but not cyanotic [blue]), and exercise intolerance or loss of stamina. This is equivalent to heat exhaustion in people.

8-51. When the symptoms of overheating occur, loosen the leash, collar, and muzzle as much as possible without loosing control of the MWD. This allows for more efficient panting and cooling. Follow these guidelines at the onset of symptoms:

- Carry the MWD to the nearest shade and fan him if no wind is present.
- Try to quickly lower the MWD's body temperature by running and sponging cool water over the MWD's head, body, and legs. If a body of water is available, allow the MWD to stand in the water but make sure the MWD's head remains above water. Running water works better than standing water.
- Place the MWD in an air-conditioned kennel, vehicle, or shelter.
- Apply ice packs to the MWD's groin, armpits, and head to increase speed of cooling.
- Dampen, but don't soak, the MWD's feet, ears, and abdomen with rubbing alcohol. Allow it to evaporate and then reapply more.
- Do not let the MWD over drink. He should not consume more than one cup of water until he is calm and his temperature returns to normal.
- Monitor the MWD's temperature every 5-10 minutes, and stop cooling efforts when it reaches 100-103° Fahrenheit. Keep monitoring his temperature for an additional 60 minutes.

8-52. If the MWD's temperature remains normal (about 100-103° Fahrenheit), treatment is complete; however the MWD must be examined by a VCO and a heat injury must be reported in his VTR. If his temperature climbs again, restart cooling efforts.

8-53. If the MWD's temperature drops below 100° Fahrenheit, begin warming by drying and wrapping him with sheets and blankets. Remove the heat when his temperature goes above 101° Fahrenheit.

8-54. For severe heat injuries, the MWD's temperature is generally over 104° Fahrenheit combined with poor response to commands, weakness, unsteady movement, vomiting, difficult or labored breathing, uncontrolled panting, convulsions, collapse, a dull or depressed attitude and behavior, a rapid but somewhat weak pulse, exercise intolerance or loss of stamina, and cyanosis (blueness) or dark red mucous membranes. This is equivalent to heatstroke in people. If this occurs take the following actions:

- Initiate all cooling and monitoring efforts as noted in *paragraph 8-51.*
- Administer 1 liter of intravenous fluids to the MWD as quickly as possible if appropriately trained personnel are available.
- Continue cooling and monitoring until a veterinarian can evaluate the MWD.

BLOAT (GASTRIC DILATION-VOLVULUS)

8-55. Bloat is an acute stomach enlargement and twisting that may be due to gas, food, or water. It may occur if the MWD is fed immediately before or after hard exercise or when he is returned to his kennel after work or exercise and allowed to drink too much water.

8-56. An enlargement may be seen just behind the ribs, primarily on the left side. The MWD may also be restless and show signs of pain in the abdominal region. He may attempt to vomit or have a bowel movement. His breathing may be difficult or labored due to pressure from the enlarged stomach.

8-57. Handlers should notify the VCO and stop all watering and feeding. Walking may enable the MWD to relieve himself through bowel movements or passing gas. Most cases require extensive treatment by the VCO.

8-58. To prevent bloat, MWDs should not be fed within a two-hour period before or after hard work or vigorous exercise. It is also safer to provide multiple small meals each day, feeding twice daily, rather than one large meal. In hot weather, give small amounts of water during training or working MWDs to prevent excessive thirst. For the first hour after working or training, only three inches of water in the bucket should be available. After this cooling-off period, more water may be given.

First Aid Kit

8-59. The responsible VCO determines the contents of MWD first aid kits. The VCO should consider the skill of the handlers, access to veterinary support, and emergency situations that may occur. The DODMWDVS does not have a standard first aid kit supply list because needs vary. The local VCO and KM are responsible for developing and putting together an appropriate kit and training the handlers how to use it. The kit should contain supplies for cleaning and dressing minor wounds, cleaning and protecting an eye injury, bandaging or splinting minor breaks, and removing mild superficial foreign bodies (splinters or glass). It should also contain peroxide, antibacterial solutions, and a small supply of any prescription medications needed for individual MWDs.

ENVIRONMENTAL EFFECTS ON THE MILITARY WORKING DOG

8-60. Just as the environment affects soldiers performing their duties, the MWD is affected by his surroundings. It is important for planners and leaders to become aware of the many different conditions that can affect the performance of MWDs. Not all MWDs are affected in the same manner. Acclimation times will vary between conditions and MWDs. Planning considerations should always include the VCO when the MWD will be used or trained in environments that he is not accustomed to.

Wind

8-61. Always consider the direction and velocity of the wind when using an MWD for any detection operations. This consideration holds true even indoors. There may be a slight breeze that will move a scent away from the location of its source. Fans, air conditioning units, and objects in motion can create winds that will affect the performance of a detector dog.

8-62. In addition to these operational concerns, wind may create medical issues (such as blowing sand) that are negative. Wind also creates increased cooling, which is positive in hot weather but negative in cold weather.

Inclement Weather

8-63. Precipitation can degrade the MWD's performance of detection operations; however, do not assume that an MWD will not be able to perform satisfactorily. The amount of moisture accumulated and the amount of time elapsed since the scent was left are variables. MWDs have been proven to be effective even in inclement weather. MWDs that become wet and are returned to their kennels or crates while wet are at increased risk of skin injury and infection.

Heat

8-64. When operating in locations of extreme heat, the MWD should be allowed sufficient time to acclimate. Consult with the VCO at the location to determine how long an MWD should be allowed to acclimate. Care should be given to ensure that the MWD does not sustain injuries to the pads of his paws due to excessive walking on concrete or pavement in extreme heat conditions. Always ensure that the MWD has a fresh and constant source of drinking water.

Cold

8-65. Extreme cold temperatures should not deter MWD operations. Special care should be taken to enable an MWD to perform in cold locations. The accumulation of moisture on the MWD's paw pads can freeze and injure him. Using paw pad protectors to aid in the prevention of injury due to cold weather may be helpful, but the MWD must be trained to wear them. The MWD should be allowed to rest and warm up on the same cycle as the handler.

TERRAIN

8-66. The safety of the handler and MWD must be given the highest consideration when operating in hazardous terrain. Care should be taken when employment will take place in rocky or mountainous areas. Unfamiliar locations can pose threats in the form of caves, holes, loose footing, and other hazards. Whenever possible, conduct a reconnaissance of the area before employing an MWD.

DISTRACTIONS

8-67. There are numerous distractions that can affect MWD operations. These are conditions that can cause the MWD to focus his attention away from the desired task at hand. Planning and consideration should be given to eliminate or reduce distractions that hinder operations. Distractions are natural and man-made. Some examples of distracters are:

- Other dogs (both MWD and indigenous).
- Other animals.
- Human presence and activities.
- Noise.
- Smells.

Appendix A
Metric Conversion Table

This appendix complies with current Army directives, which state that the metric system will be incorporated into all new publications. *Table A-1* is a conversion chart.

Table A-1. Metric Conversion Chart

US Units	***Multiplied By***	***Equals Metric Units***
	Length	
Feet	0.30480	Meters
Inches	2.54000	Centimeters
Inches	0.02540	Meters
Inches	25.40010	Millimeters
Miles (nautical)	1.85320	Kilometers
Miles (statute)	1.60930	Kilometers
	Area	
Square inches	6.45160	Square centimeters
Square feet	0.09290	Square meters
	Volume	
Cubic feet	0.02830	Cubic meters
Fluid ounces	29.57300	Milliliters
	Temperature	
Degrees Fahrenheit	Subtract 32, multiply by 5/9	Degrees Celsius
	Weight	
Ounces	28.34900	Grams
Pounds	0.45359	Kilograms
Metric Units	***Multiplied By***	***Equals US Units***
	Length	
Centimeters	0.39370	Inches
Kilometers	0.53960	Miles (nautical)
Kilometers	0.62137	Miles (statute)
Meters	3.28080	Feet
Meters	39.37000	Inches
Millimeters	0.03937	Inches
	Area	
Square centimeters	0.15500	Square inches
Square meters	10.76400	Square feet
	Volume	
Cubic meters	35.31440	Cubic feet
Milliliters	0.03380	Fluid ounces

Table A-1. Metric Conversion Chart (Continued)

Metric Units	***Multiplied By***	***Equals US Units***
	Temperature	
Degrees Celsius	Multiply by 9/5, add 32	Degrees Fahrenheit
	Weight	
Grams	0.03527	Ounces
Kilograms	2.20460	Pounds

Appendix B

Kennel Construction

This appendix is intended to be a starting point in the design process with the identification of requirements and spaces. It should serve as a supplement to the installation design guide and not as a replacement. This appendix provides commanders and planners with guidelines and specifications to consider for the construction and equipment of kennel facilities. The floor plan (*Figure B-1*) is an example of how an MWD facility works, how areas relate to each other, and what is required for each space. It is not intended to serve as a rigid model for duplication.

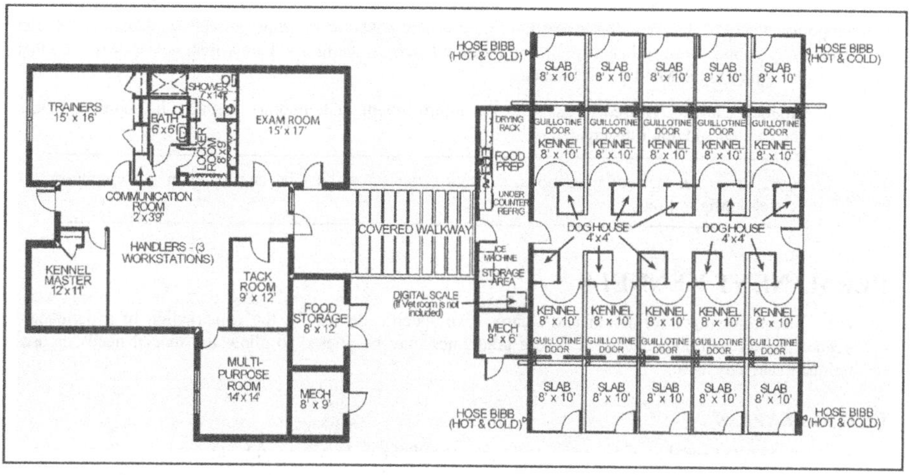

Figure B-1. Example of an MWD Kennel Floor Plan

TEMPORARY KENNELS

B-1. There are two major types of kennels, temporary (*Figure B-2*, page B-2) and permanent. According to *AR 190-12*, temporary kennels are to be used during deployments for no longer than one year, to allow sufficient time for permanent kennels to be built. Temporary kennels may be constructed out of any available resources. The design is not standard and varies from site to site depending on the location, mission, and duration of operations. Using a concept similar to constructing fighting positions, the temporary kennels will be improved continuously as needed or until a permanent facility is completed.

B-2. Using tents and shipping crates may be a hasty method of setting up temporary kennels. Although this method is not optimal for sustained operations, it may be useful until other accommodations can be coordinated. Erect the tents at a location that allows the MWDs a safe place to rest away from other troop activities after operations. Whenever possible, use terrain or buildings to place a barrier between troop activity and the temporary kennel. Never place the temporary kennel site near motor pools, dining facilities, or high noise areas.

Figure B-2. Temporary Kennel

B-3. In some areas of operation, it may be possible to obtain the use of a building or other structures for temporary kennels. In these cases, leaders must ensure that the building is well ventilated, safe, and structurally sound. Ensure that there are no hazards (such as chemicals, bare wires, holes, or debris) that can cause injury to the MWDs or handlers.

B-4. Engineers are often used to assist in the improvement of temporary kennels. In addition, fences should be erected and dog runs created.

> ***Note*:** From the start of operations, sanitation measures must be employed in all temporary kennel configurations.

PERMANENT KENNELS

B-5. The following information and figures give specific details for the construction of a permanent kennel facility and adjacent areas. These guidelines may be altered to allow for special needs or uses unique to each location.

EXTERIOR AREAS

B-6. An MWD kennel complex (*Figure B-3*) consists of kennels, a support building, an obedience course, an exercise area, an MWD break area, exterior storage, parking areas, drives, and walks. Often, as a result of the location, utility services are limited at the site. Construction should provide the following utilities:

- Water (potable with backflow prevention).
- A sanitary sewer.
- Electric.
- Telephone.
- Network and communication service to the site.

Other systems should be incorporated according to local procedures.

B-7. The entire complex should be enclosed, at the minimum, with a heavy-duty, 8-foot-high chain-link fence with three strands of straight wire (no barbed) at the top to prevent an MWD from climbing or jumping out. A 10-foot (minimum) vehicle gate that can be padlocked is to be installed to allow for food deliveries to the kennel and other access requirements. A personnel entry gate that is visible from the KM's office is also required. The personnel gate should have a cipher lock and a notification buzzer that sounds in the KM's office, the obedience course, and the kennel. The buzzer in the kennel should have the capability to be turned on when personnel are in the area and turned off to allow the MWDs to rest when personnel are absent. All gates should be self-closing and self-latching.

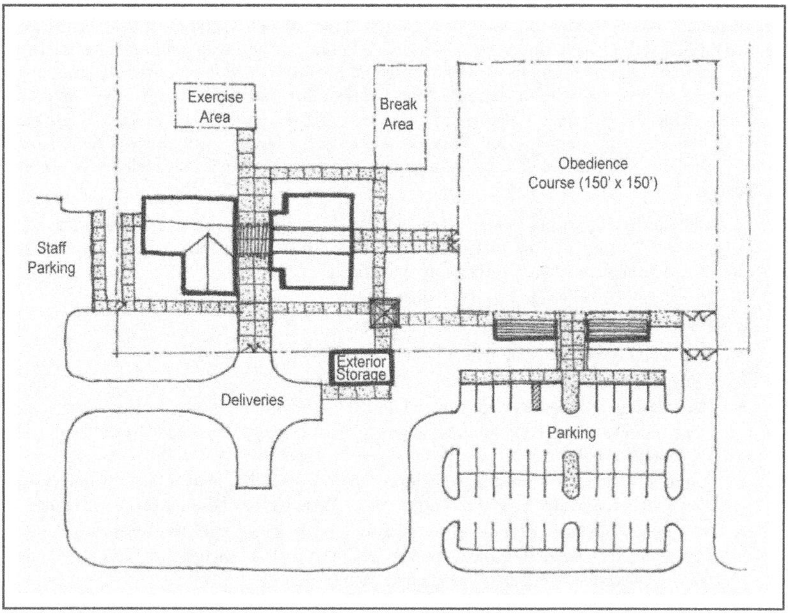

Figure B-3. MWD Kennel Complex

B-8. Security lighting for the area should be according to local policy. At a minimum, the lighting should be on from dusk to dawn and placed on all corners of the fencing area. Security lights should be placed in areas that the other lights do not cover.

B-9. Warning signs will be posted on the exterior fencing and buildings of the MWD kennel and exercise area. Signs should contain the following words: "DANGER—OFF LIMITS–MILITARY DOG AREA." Personnel approaching the kennel area should be able to see and read the warning signs under normal daylight conditions from a distance of 50 meters.

B-10. Locate a trash dumpster for the complex outside of the perimeter fence. Screen the dumpster and provide low-maintenance landscaping for the complex according to local policy.

Exterior Storage, Parking Areas, Drives, and Walks

B-11. An enclosed, 400-square-foot exterior storage building is required to store maintenance equipment, portable kennels, and obstacles. Interior/exterior lighting, power, and a water faucet should be included.

B-12. Staff parking should be a paved parking lot large enough for eight POVs, three GOVs, two visitor vehicles, and one handicap parking space. Pave access drives to accommodate parking, exterior storage, food deliveries, emergency vehicles, and the transportation of MWDs.

B-13. Provide paved walks to all areas. When designing, give careful consideration to creating a one-way MWD traffic system. To prevent confrontations between the MWDs, avoid situations that allow MWDs to meet head on.

Obedience Course (150 Feet by 150 Feet)

B-14. The obedience course plays an important role in maintaining the MWD's agility and stamina as well as in reinforcing obedience and proficiency training. The course should be grassed and free of hazards (trees, large rocks, holes, and burrs) that may be harmful to MWDs and handlers. The site should be graded for drainage but minimally sloped to provide a level field for training. The area should be enclosed with an 8-foot chain-link fence. Gates should be self-closing and self-latching with a minimum width of 5 feet to allow for lawn maintenance equipment. The area should be well lighted to eliminate shadows. Light poles should be located on the exterior of the fence. On and off switches should be manual. Weatherproof outlets should be provided at each light to provide power at the site. Speaker units connected to the chosen public address system should also be provided. Water should be at the course site with water faucets installed to provide water for MWDs and lawn watering. An irrigation system is desirable for long-term lawn maintenance.

B-15. Obstacles for the course (*Figure B-4*) include barrels, tunnels, steps, jumps, framed windows, and dog walks. Each obstacle is 15 to 20 feet from the previous obstacle; the course is run in a sequence. Obstacles should be constructed to the following specifications:

- Barrel 1 is 35 inches long with a 23-inch opening.
- Barrel 2 is 70 inches long with a 23-inch opening.
- Barrel 3 is 105 inches long with a 23-inch opening.
- The tunnel is 146 inches long with a 19-inch opening.
- The steps are constructed out of wood and are 94 inches high with 5 steps on each side. Each step is 43 inches wide by 16 inches high by 24 inches long. The top platform is 48 inches long by 43 inches wide.
- Jumps 1, 2, and 3 are constructed of wood and are 36 inches high at the maximum raised level. There are 6 removable boards that are 51 inches long by 6 inches high and 1 inch thick.
- The framed window is constructed of wood and is 8 feet tall. The maximum height for the MWD to negotiate is 36 inches. The window opening is 48 inches wide. Removable boards are 5 1/7 inches wide by 51 inches long and 1 inch thick.
- The A-frame is constructed of wood. Its two sides are 8 feet long by 4 feet wide by 1 inch thick with 1-inch by 2-inch boards every 6 inches to secure footing. It has a maximum raised height of 6 feet with a base spread of 10 feet.
- The dog walk is constructed of wood and is 26 inches high by 224 inches long and 11 inches wide with 1-inch by 2-inch boards every 4 inches to secure footing. Each end has a 45° incline ramp that is 11 inches wide by 49 inches long (included in the 224 overall length measurement).

Note: See *DA Pam 190-12* for more complete construction guidelines including the bracing necessary for the various obstacles.

Military Working Dog Break Area (10 Feet by 20 Feet)

B-16. An MWD break area should be located near the kennels. A break area allows the handler to release the MWD immediately after exiting the kennels or before entering the kennels so that the MWD can relieve himself. The area should be enclosed with an 8-foot chain-link fence. Gates should be self-closing and self-latching and a minimum of 5 feet wide to allow for lawn maintenance equipment. The ground can be a sandy area for cleaning ease. A water faucet is required to maintain the area.

Exercise Area (20 Feet by 40 Feet)

B-17. The exercise area is a space where the MWD can be released without the handler being present. The exercise area should not be a shared area with the obedience course, as it would conflict with training objectives. The exercise area should be visually separated from the obedience course to prevent MWD from being distracted during training. The exercise area should be grassed, hazard free, and graded slightly for drainage purposes.

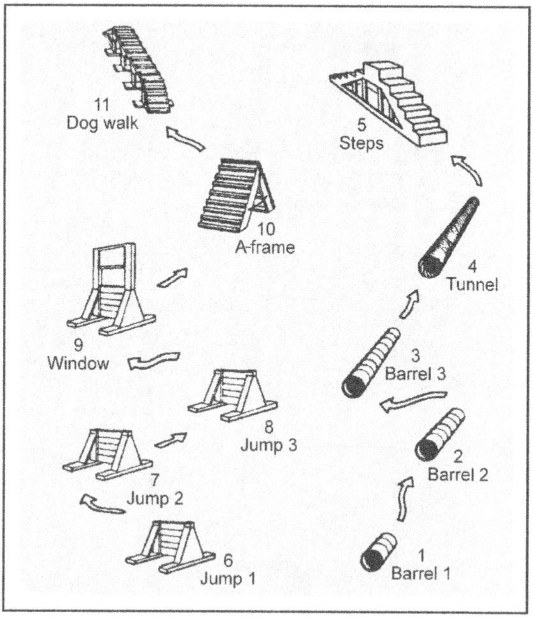

Figure B-4. Obstacle Course Layout

INTERIOR AREAS

B-18. Interior areas have heating, ventilation, and exhaust systems; general-purpose lighting; and power outlets in all spaces. Provide phone, fax, and local area network (LAN) connections to support each area and hot and cold water to all latrines, sinks, and showers. Construct the facility to meet all fire protection and life safety feature requirements.

Administration Areas

B-19. Since MWDs are likely to visit the administration area (*Figure B-5*, page B-6) at some point, finishes should be of durable material, require minimal maintenance, and be easily repaired or replaced in case of damage. (For example, flooring should be covered by carpet.)

Kennel Master's Office (132 Square Feet)

B-20. The KM's office serves as the nerve center for the complex. Ideally, the office should be located at the front of the administrative area with exterior windows that view the entry gate to the complex. It includes a lockable closet for supply storage. Furnishings include a desk, a chair, two visitor chairs, a four-drawer file cabinet, and a bookcase. Equipment includes a computer, a printer, a fax/copier, and a public address system or central intercom console.

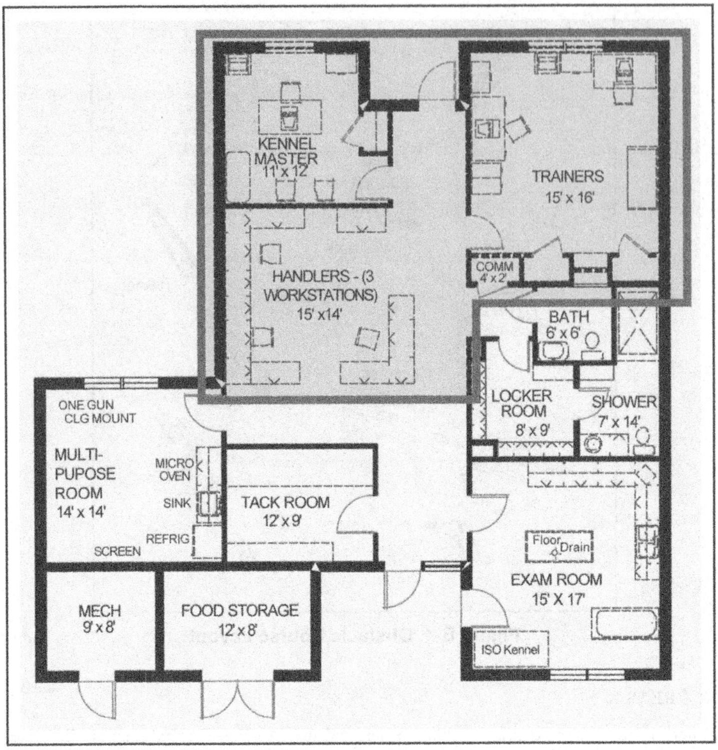

Figure B-5. Kennel Administrative Areas

Trainer's and Plans Noncommissioned Officer's Office (240 Square Feet)

B-21. The trainer office should provide workspace for two people. Lockable closets are required to store training aides for each trainer. Furnishings include a desk, a chair, and a four-drawer file cabinet for each trainer and a common bookcase. Equipment includes a computer, a printer, a copier, and a 4- by 8-foot dry erase board.

Handlers' Area (210 Square Feet)

B-22. The area for the handlers is an open area for general office functions. It includes—

- Three common workstations, depending on the size of the kennels, for handlers to update records and complete daily reports.
- Space for document storage.
- Furnishings (three workstations, file cabinets [one four-drawer vertical file per workstation or the equivalent], and a distribution counter.
- Equipment (computers and a printer).

As a general rule, one workstation should be available for every three handlers assigned to an MWD section.

Controlled-Substance Storage Room

B-23. This room is inside the administrative area and allows sufficient room for the storage safe in which controlled-substance training aids are stored. The doors are secured using approved locking devices. Lightweight doors are replaced with solid metal or wood doors or covered with a 9- to 12-gauge security screen or 16-gauge sheet steel. They are fastened with smooth-headed bolts and nuts and peened in place. All windows providing access to a storage room that is not staffed 24 hours a day are protected by a 9- to 12-gauge security screen or 3/8-inch or larger diameter steel bars spaced no more than 6 inches apart. The frames holding the screen or bars must be fastened to the window frame with smooth-headed bolts.

Special-Use Areas

B-24. Special-use areas (*Figure B-6,* page B-8) are the transition spaces between the administration area and the kennels. They are important in the day-to-day operations of the MWD facilities. They are not normally occupied on a daily basis. As such, these areas provide a secondary function of serving as a noise buffer between the kennels and the administrative areas. The special-use areas provide a workspace for the VCO and accommodate the unique storage requirements for the kennels.

Veterinary Treatment Room

B-25. A veterinary treatment room is used to perform health care and first aid for MWDs since many locations do not have veterinary treatment areas in close proximity to the kennels. In order to maintain sanitary conditions, treatment rooms require a higher degree of cleanliness and durability than the rest of the support building. Basic finish requirements are—

- Seamless floors with an integral base.
- Washable walls.
- Painted gypsum board ceilings.
- Cabinets faced with plastic laminate.
- Solid-surface countertops.

B-26. Interior partitions should provide a sound transmission class rating of 50 to 55. Floor drains are centered in the room to aid in cleaning, and a dedicated floor drain is located under the isolation kennel. Wall and base cabinets with drawers are installed, including stainless steel dual washbasins with hot and cold water. One section of wall cabinets should be lockable so that prescription medicines for MWDs may be stored in the room. In addition to the normal electrical outlets required by code, a 220-volt outlet is required to support other portable equipment that may be used. Equipment for the room includes the following:

- **An MWD isolation kennel.** The entire structure should be made of uniform, nonglare type 304 stainless steel. The frame, cross members, and vertical spokes should be welded together at every intersection. Latches and hinges are made of 12-gauge stainless steel. The floor of the pen is a raised, removable grate. The floor grate is a heavy-gauge stainless steel, plastic-coated section installed 1 3/4 inches off the floor. A dedicated floor drain is required for this kennel.
- **Standard table tub**. The tub is made of type 304 stainless steel with a minimum depth of 15 inches. The tub slopes to the drain end. The design includes a removable, bathing rack made of plastic-coated stainless steel that fits into the bottom of the tub to give MWDs better footing during treatments. Faucets include a flexible, stainless steel hose that will extend the length of the tub and spray powerfully to reach all areas of the tub.
- **Exam room light.** A ceiling-mounted exam room light with an arm extension and swivel capacity of 360° is recommended.
- **Walk-on platform scale.** The scale is a type 304 stainless steel platform with a wall- or post-mounted light-emitting diode (LED) display. The scale is able to perform calibration and weigh animals to the nearest 0.1 of a pound or kilogram.
- **Stationary exam table.** The table top is type 304 stainless steel die-formed from one piece of 20-gauge steel. It should have raised edges to prevent fluid runoff and be permanently attached

to a heavy base. There should be a minimum 35-inch work height from the floor to the top of the table. The heavy base should have adjustable leveling screws.

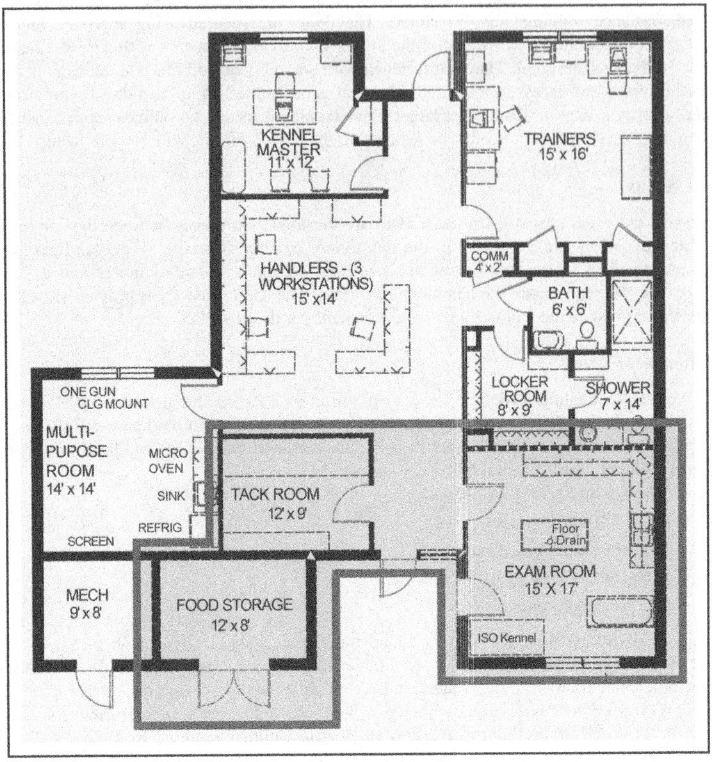

Figure B-6. Kennel Special-Use Areas

Tack Room (108 Square Feet)

B-27. The tack room is for the storage of extra MWD equipment, such as bite suits and portable kennels. The tack room is located near the entrance to the kennels but not the main building entrance. Requirements include 12 linear feet of shelves that are 18-inches wide and 36 inches deep and allow for 36 inches above each shelf; a 24-inch space with hooks to hang items on; and 6 linear feet of shelves that are 18 inches wide and 12 inches deep and allow for 18 inches above each shelf.

Food Storage Room (96 Square Feet)

B-28. The food storage room is normally for the bulk storage of a 30-day supply of food for the MWDs. The room is adjacent to the kennels with an exterior entry. The entry is a pair of 36-inch doors to allow for palletized food delivery and storage. Doors require weather stripping and thresholds to aid in insect and rodent control. Temperature and humidity control equivalent to the office area is required to control food spoilage. A minimum separation of 2 inches is required between the walls and the stored food to allow for air circulation. Shelving should be installed to meet local delivery operations.

SUPPORT AREAS

B-29. The support areas (*Figure B-7*) provide space for the common-use areas of the building. Areas for the multipurpose room, latrines, communications, and mechanical necessities form the core of the support areas. Support areas should be centrally located and easily accessible for building occupants or maintenance personnel in the case of the mechanical room.

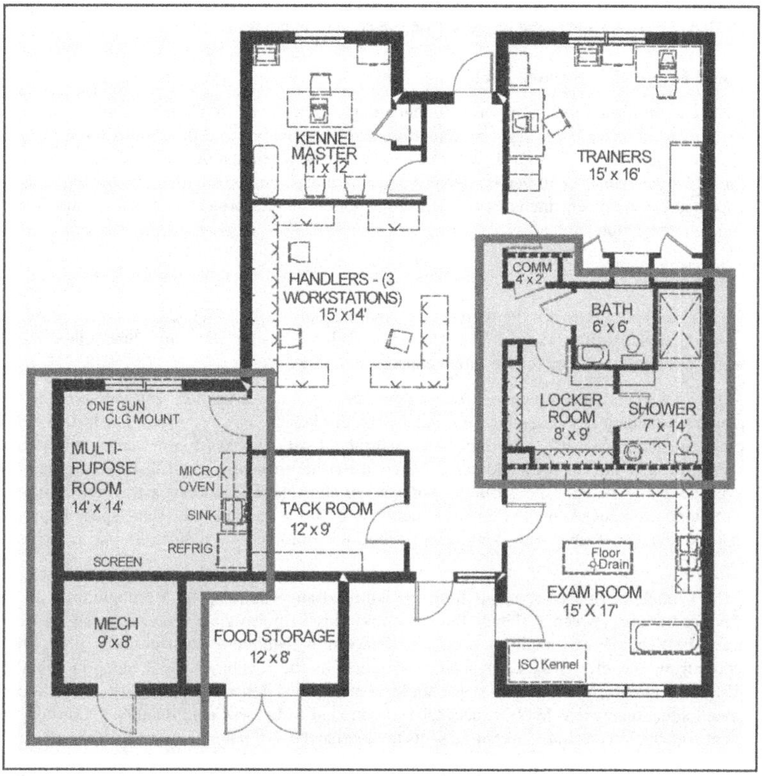

Figure B-7. Kennel Support Areas

Latrine, Shower, and Lockers (206 Square Feet)

B-30. A unisex area is divided into three rooms for the latrine, shower, and lockers. One room is a handicapped accessible latrine with a water closet and lavatory for general use. The locker room includes 10 full-height metal lockers and a bench for MWD handlers' use. The shower room includes a shower, bench, water closet, and lavatory. Do not neglect to include other accessories such as mirrors, soap dispensers, paper towel dispensers, disposal units, grab bars, and hooks.

Heating, Ventilation, Air Conditioning, Mechanical, Electrical, and Communications Systems

B-31. Heating, ventilation, and air conditioning (HVAC) is a process to add hot or cold air in a facility. It also introduces fresh air in to the facility to maintain indoor air quality and eliminate sick building syndrome. Design HVAC systems to optimize energy use and meet the design criteria for the area in which you are located. Place the hot water heater in the mechanical room and with a capacity that will meet the demands of the latrine and shower; multipurpose room; and veterinary treatment room. Locate electrical panels in this area as well. Although a recommended size is shown on the plans, the room should be big enough to support the equipment required for the location.

B-32. While the number of instruments and computers is not large, the communications closet is vital to the success of the organization. Provide a minimum of one prewired telephone and LAN outlet for each workstation, one in each common area, and one in the kennel food preparation area. Include telephone and LAN outlets for fax machines and printers. Conceal conduit in the walls.

Multipurpose Room (196 Square Feet)

B-33. The multipurpose room serves as a combination conference room, classroom, and break room. Food preparation requirements are a light, a countertop, and base and upper wall cabinets for storage. Lighting should have a variable-level control for use as a conference room and/or classroom. Furnishings include either a conference table for fifteen or three five-person tables; chairs; a bulletin board; and a 4- by 8-foot dry erase board. Equipment includes a video projection system, a projector screen, a sink, a microwave oven, and a refrigerator.

KENNEL AREA

B-34. The kennel area provides the daily living environment for the MWDs, giving each MWD a private place for eating and resting. Kennel areas should be built to accommodate large-breed dogs and designed as modular structures to allow for the future expansion of runs.

B-35. There are three types of kennel areas; indoor kennels, outdoor kennels, and indoor/outdoor kennels. The preferred standard is the combination indoor/outdoor kennel; however, there are factors that must be taken into consideration when selecting the kennel type. Cost is always a consideration, but climate also influences the selection. In cold weather climates, the benefit-to-cost ratio will drive the selection to an indoor kennel while the opposite is true for hot weather climates. The more temperate the climate the more suited the indoor/outdoor kennel is for use. Each type of kennel includes the same space requirements. A kennel includes kennel runs with doghouses, a food preparation area, a storage area, and a mechanical room.

B-36. The kennel should be separated from the administrative building by a minimum of 20 feet. This provides separation between different functions and aids in noise control. As a result of the constant interaction between the two facilities, a covered walkway is required to connect the facilities and provide protection from the elements. Care should be taken in the layout of the kennel in relation to the administrative building; a one-way system should be maintained in the movement of MWDs. The objective is to avoid situations where MWDs meet face to face and to prevent confrontations. Consider installing mirrors at corners, intersections, and blind spots to alert handlers of potential wrong-way traffic.

B-37. All entries and exits from the kennel area are required to be self-closing and self-latching. Doors exiting the kennel runs should open inward to aid in preventing MWD escape. Proper ventilation is important in the kennels to prevent the spread of diseases and to control odors. It is recommended that the ventilation standard be to cycle out the room air 10 to 15 times per hour. Kennel temperature should range from 45° to 85° Fahrenheit with humidity in the range of 40 percent to 70 percent. Temperature should be maintained within 10° of the exterior temperature. MWDs work more effectively and are more alert when the kennel temperature is close to the temperature of their working environment.

B-38. Potable water is required at the kennels. Hot and cold water lines are installed at ceiling level down the length of the 5-foot wide center corridor connecting to hose reels at each end of the corridor. A high-pressure washer is required to assist in sanitizing and cleaning kennel runs.

B-39. Since the kennel is basically a wet environment, all electrical receptacles should be provided with ground fault circuit interrupters and all-weather covers. Receptacles mounted at 48 inches above the finished floor are required at each end of the central corridor.

B-40. Each kennel run should have a waterproof light fixture that has an individual switch at the corridor entrance. Other areas require general-purpose lighting according to industry standards. All drain lines in the kennel should be a minimum of 6 inches in diameter and be designed to sustain flow velocities that will maintain a self-cleansing action. Due to the nature of the waste, cleanouts should be numerous and easily accessible. Floor drains should be included in the central corridor to aid in the cleaning of kennels.

Indoor/Outdoor Kennel

B-41. A combination indoor/outdoor kennel (*Figure B-8*, page B-12) simply implies that there is an interior run and an exterior run linked with each MWD living area. A guillotine type door will connect the interior run and the exterior run. Other door types may be used, but the idea is to permit isolation of the MWD during cleaning operations. Ideally, operation of the door is from the central corridor and from the outside of the exterior run. At 6 feet above the finished floor, use translucent wall panels in the wall separating the runs to provide natural light inside the kennel.

B-42. The roof should extend over the exterior runs and include gutters and/or downspouts to prevent excessive water in the runs. Exterior framing should be enclosed with a soffit to prevent birds from roosting. Exterior water faucets (with freeze protection where required) with both hot and cold water are required at each corner of the building. These are necessary for cleaning exterior runs. Floor drains (*Figure B-9*, page B-12) should be located in the corner of each run adjacent to the common wall separating exterior and interior runs. Drains should tie-in to a common waste line. Drain covers should be flush with the floor. Kennel floors should slope (at least 1/4- inch per foot) toward the drains to allow for quick water drainage and drying.

Indoor or Outdoor Kennel

B-43. The plan for the indoor or outdoor kennel (*Figure B-10*, page B-13) is the same for either style. The obvious difference is the wall that encloses the kennels. For indoor kennels, the exterior walls along the runs should have windows beginning at 6 feet above the finished floor to allow for natural light. Windows should be operable and hinged at the sill to tilt inward to prevent the escape of MWDs. For outdoor kennels, the exterior wall is simply an 8-foot chain-link fence. Water faucets with hot and cold water are required at each corner of the building. These are required to clean the gutter and drain system (*Figure B-11*, page B-13) for the kennel. A gutter and drainage system is provided at the back of the kennel run for either type of kennel for sluicing wastewaters from cleaning operations. Kennel run floors should slope to the gutter and the gutter should slope to the drain. Gutters should be sufficient to allow for easy cleaning. A minimum of 3 feet should be allowed between the end of the kennel and the wall or fence.

Kennel Run

B-44. Runs should be arranged so that openings are staggered in order to avoid MWDs facing each other across the corridor. Full-height partitions are required between runs for all kennel types. The first 6 feet of wall above the floor should be sealed concrete. For interior runs, steel clad acoustical panels should be considered from the top of the concrete partition to the ceiling. Steel clad acoustical ceiling panels should also be considered above the kennels.

B-45. Heavy-gauge chain-link fence is used for exterior runs and is also an option for interior runs. Because of the noise generated in the enclosed area of an interior run, it is important that noise-reducing materials be introduced to prevent stress to the MWD and hearing loss to personnel.

B-46. As a minimum, end walls should be full-height, galvanized chain-link fencing. Take care to prevent sharp edges from bolts, hinges, and tie wires on the interior of the runs where MWDs may be cut or harmed. Self-latching entry gates should swing 180° and be able to cover the opening of the MWD house. This allows the MWD to be penned in the house during run-cleaning operations. A stainless steel water bucket is required for each kennel, including a holder that prevents the MWD from overturning the bucket.

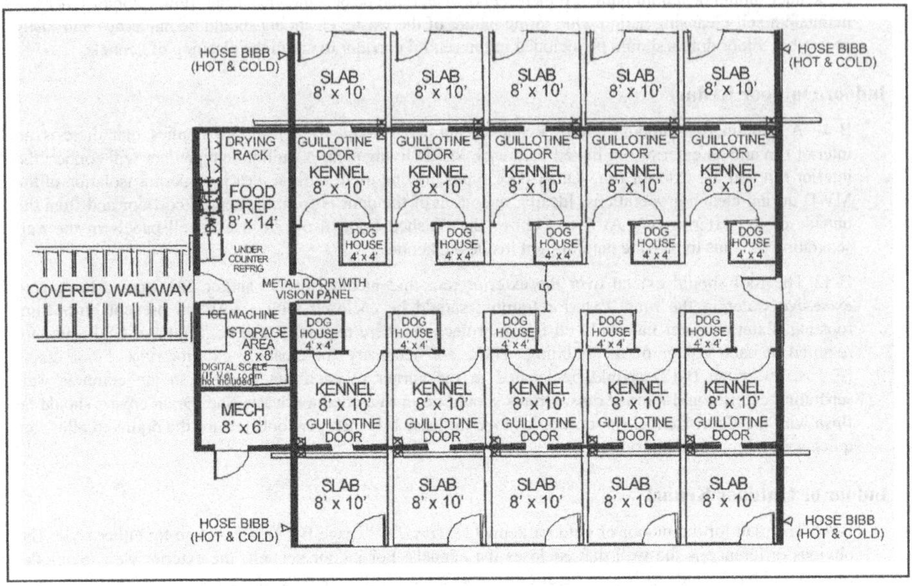

Figure B-8. Combination Indoor/Outdoor Kennel

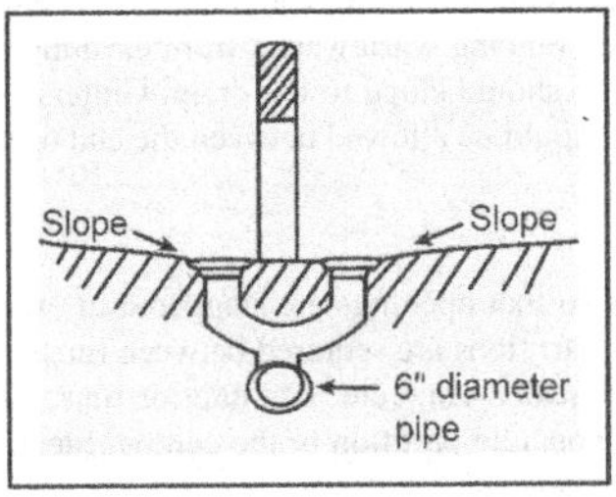

Figure B-9. Example of a Combination Indoor/Outdoor Kennel Floor Drain

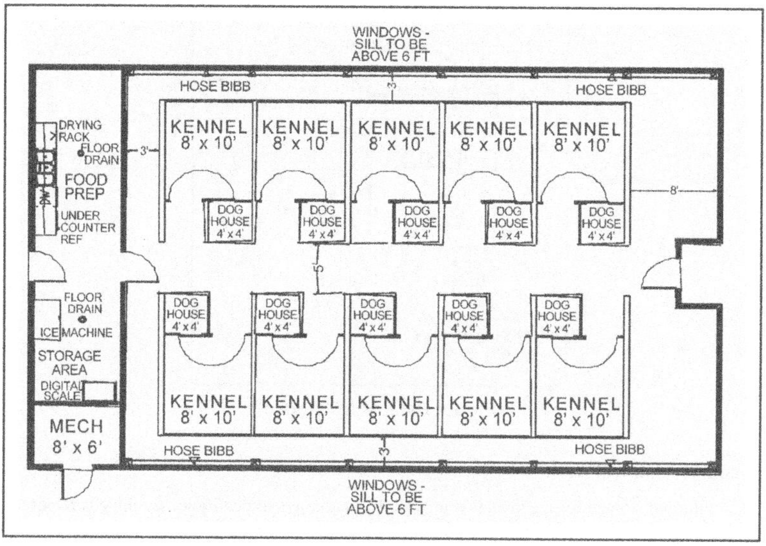

Figure B-10. Indoor or Outdoor Kennel

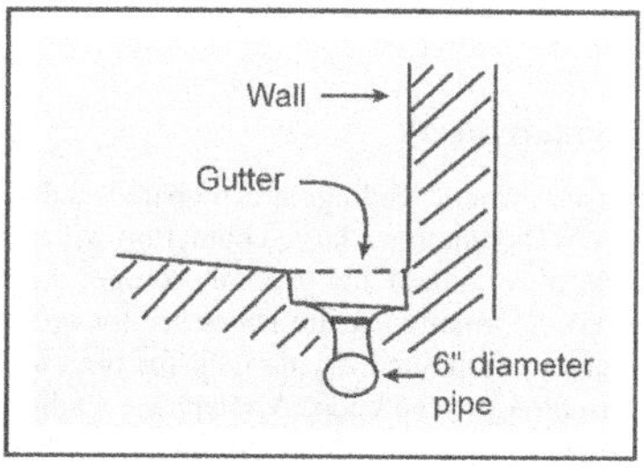

Figure B-11. Example of an Indoor or Outdoor Kennel Floor Drain

B-47. Flooring should be concrete and slope toward floor drains and/or gutters to provide rapid runoff of water and drying. Since runs are subject to frequent wash downs to clean and prevent disease, a smooth epoxy-type finish is needed. Temperature-controlled floors should be emplaced for locations that have inclement weather.

B-48. All materials in areas accessible to MWDs should be resistant to damage by scratching, biting, and chewing. Materials should be durable and easy to sanitize. Avoid using material such as angle iron that rusts over a period of time and generates sharp edges that pose a danger to MWDs.

B-49. MWD houses are 4 feet by 4 feet by 4 feet sealed, concrete boxes with a metal top. Tops should hinge toward the end wall in order to be raised to clean the house or treat an MWD. Once lifted, a holding device should be available to keep the top open. Wall opening dimensions are shown in *Figure B-12*, page B-14. The house should be a minimum of 8 inches and a maximum of 12 inches above the run floor. The house floor should have positive drainage toward the wall opening.

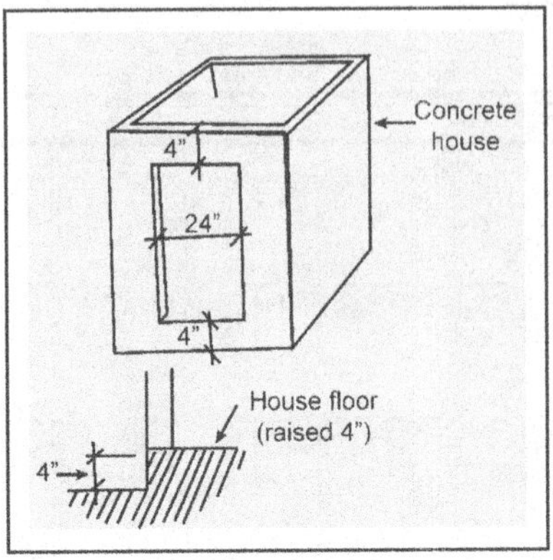

Figure B-12. MWD House

Storage and Mechanical Area

B-50. Additional storage for the kennel is provided. A commercial quality ice machine is required to aid in cooling down MWDs. Include a floor drain for the area. If a veterinary treatment room is not included, a digital scale should be located in this area. Design mechanical systems to optimize energy use and meet the design criteria for the area where the kennel is located. Include the hot water heater in the mechanical room and with a capacity to meet the demands of the kennel. Allow for the location of electrical panels in this area as well.

Food Preparation Room (112 Square Feet)

B-51. The food preparation area is where feedings are prepared and feeding pans/bowls are stored and cleaned. Food preparation for MWDs requires a large countertop with a deep, three-compartment stainless steel sink. Upper wall cabinets are required for general storage. An under the counter refrigerator, a dishwasher, and space to store food for daily use are necessary for operation. A 2-foot space at the end of the counter should accommodate a storing and drying rack for pans and bowls. Lighting and receptacles and a floor drain should be installed per the code. A telephone, bulletin board, feeding chart, and large garbage can should also be provided. Optional equipment includes a garbage disposal unit and a stackable washer and dryer.

Appendix C

Inspection Memorandum

This appendix provides a sample memorandum (see *Table C-1*, page C-2) for documenting the results of inspections. Commanders and program managers will find this useful in preparing for inspections. KMs, plans NCOs, and handlers will find this list helpful in conducting daily duties and adhering to requirements for inspections. Maintain completed memorandum records on file to assist in future inspections.

Table C-1. Sample Inspection Memorandum

DEPARTMENT OF THE ARMY
UNIT
ADDRESS OF UNIT

OFFICE SYMBOL 15 May 2004

MEMORANDUM FOR RECORD

SUBJECT: MACOM Technical Assistance Visit or Provost Marshal Inspection

1. SECTION I. GENERAL INFORMATION.

UNIT: 529th Military Police Company LOCATION: Kaiserslautern
DATE: 3-14-05 TIME: 0900

INSPECTOR (S): LTC Jeffery S. Captain ESCORT: SFC James Majors

TYPE OF VISIT: __X__ ANNOUNCED______UNANNOUNCED

TYPE	AUTHORIZED	ON HAND
Patrol		
Patrol Explosive	6	4
Patrol Narcotic	3	2
Explosive Detector	0	0
Narcotic Detector	0	0

2. SECTION II. PERSONNEL.	**YES**	**N/A**	**NO**
a) Are all authorized MWDs on hand (unit TDA/MTOE document)?	____	____	____
b) Are all MWD handlers certified by the DOD Dog Center?	____	____	____
c) Are ASI trained handlers being used as specified in AR 614-200?	____	____	____
d) Is the assignment policy of one dog and one handler being followed?	____	____	____
e) Are all MWDs assigned to a handler?	____	____	____
f) Has an excess of dogs been reported to the MACOM?	____	____	____
g) Does the KM meet the minimum criteria to be appointed as a KM?	____	____	____
h) Are handlers being assigned duties that take away from their handling responsibilities (excluding PT, staff duty, field duty, and unit training)?	____	____	____
3. RESPONSIBILITIES OF THE PROVOST MARSHAL AND COMMANDER.			
a) Is there an updated SOP signed by the PM explaining policies, procedures, and responsibilities for the direction, management, and control of the kennels?	____	____	____
b) Has the PM initiated an MWD program based on the evaluation of the installation support, customs, HRP support, and combat support missions?	____	____	____

Table C-1. Sample Inspection Memorandum (Continued)

OFFICE SYMBOL SUBJECT: MACOM Technical Assistance Visit or Provost Marshal Inspection	YES	N/A	NO
c) Has the PM issued written guidelines governing the release of an MWD to apprehend a suspect?	___	___	___
d) Are there specific criteria in contingency plans governing the use of dog teams during civil disturbance situations?	___	___	___
e) Does the PM provide guidance on procedures for explosive detector dog support to civil authorities?	___	___	___
f) Is the proficiency of each MWD team evaluated periodically by the PM?	___	___	___
g) Has the PM and/or unit commander checked one or more of these areas monthly: kennels, kennel support building, training and exercise areas, equipment, handlers, dogs, training, utilization and team proficiency?	___	___	___
h) Has the PM and/or unit commander inspected all areas of the MWD kennels quarterly, and is a written record on file?	___	___	___
i) Are MWD requisitions submitted according to *AR 190-12* and *AR 700-81*?	___	___	___
j) Are MWDs placed on the unit property book by NSN, name, and tattoo number?	___	___	___
k) Are MWDs assigned to the KM on a *DA Form 3161* (*Request for Issue or Turn-In)* hand receipt (according to *DA Pam 710-2-1*)?	___	___	___
4. VCO RESPONSIBILITIES.			
a) Are the dogs examined by the VCO every six months (according to *AR 40-905*)?	___	___	___
b) Does the VCO inspect the kennel facility and area for proper sanitary standards on a quarterly basis?	___	___	___
c) Is the issued dog food and diet checked periodically?	___	___	___
d) Is there a program in effect for insect and rodent inspection?	___	___	___
e) Does the VCO concur with the kennel standards/facilities?	___	___	___
If not what course of action has been taken? ______________________________			
f) Does the VCO provide regular training to handlers on MWD health care, feeding, and first aid?	___	___	___
g) Does the VCO provide a list to the KM concerning the deployable status of all MWD?	___	___	___
h) Does the VCO issue health certificates for dogs pending shipment?	___	___	___

Table C-1. Sample Inspection Memorandum (Continued)

OFFICE SYMBOL
SUBJECT: MACOM Technical Assistance Visit or Provost Marshal Inspection

		YES	N/A	NO
5.	**WARNING SIGNS.**			
a)	Are there warning signs with the following words posted at each instillation entrance, "**CAUTION—THIS AREA PATROLLED BY MILITARY WORKING DOGS**?"	___	___	___
b)	Are kennels located away from disturbing influences and adequately posted with "**DANGER—OFF LIMITS—MILITARY DOG AREA**?"	___	___	___
c)	Are the shipping crates stenciled with "**CAUTION—MILITARY WORKING DOG—DO NOT TAMPER WITH ANIMAL?"**	___	___	___
d)	Are shipping crates stenciled with "**LIVE ANIMAL**?"	___	___	___
e)	Are removable or permanent markings placed on the sides and rear of vehicles with the following phrase "**CAUTION—MILITARY WORKING DOGS**?"	___	___	___
6.	**KENNEL AREA.**			
a)	Has a risk analysis been conducted on the kennels?	___	___	___
b)	Is the kennel manned both day and night?	___	___	___
c)	Does the kennel have an IDS/CCTV system if not manned both day and night?	___	___	___
d)	Are approved locks and padlocks used to secure government equipment?	___	___	___
e)	Have primary and alternate key custodians been appointed, in writing, by the PM or commander?	___	___	___
f)	Are personnel authorized to issue and receive keys listed on an access roster?	___	___	___
g)	Are all keys controlled and accounted for on a *DA Form 5513-R* (*Key Control Register and Inventory)?*	___	___	___
h)	Is the *DA Form 5513-R* secured in an approved, lockable container?	___	___	___
i)	Is the key depository box permanently affixed to a wall?	___	___	___
j)	Are keys inventoried semiannually, and is the last report on file?	___	___	___
k)	Are all keys inscribed with a serial number?	___	___	___
l)	Are kennels used to house or care for other animals?	___	___	___

Table C-1. Sample Inspection Memorandum (Continued)

OFFICE SYMBOL SUBJECT: MACOM Technical Assistance Visit or Provost Marshal Inspection	YES	N/A	NO
m) Are adequate training areas adjacent to the kennel areas?	____	____	____
1) Is the training area fence 8 feet high?	____	____	____
2) Does the overhanging fence face inward?	____	____	____
3) Is the fence staked to the ground?	____	____	____
n) Does the kennel area have an adequate water supply?	____	____	____
o) Are the kennels of a standard design?	____	____	____
7. EQUIPMENT.			
a) Is all authorized equipment on hand?	____	____	____
b) Are Air Force shipping crates returned within 10 days after receiving a new or replacement MWD?	____	____	____
c) Is there one shipping crate available for each dog authorized?	____	____	____
d) Is leather equipment properly maintained and cleaned?	____	____	____
e) Is all metal equipment free of rust?	____	____	____
f) Is the web equipment cleaned with proper cleaning material to prevent deterioration?	____	____	____
g) Are there adequate medical supplies on hand for first aid treatment and training?	____	____	____
8. USE OF MWDs.			
a) Are the dog teams being employed a minimum of 24 hours per week?	____	____	____
b) Are PDs employed in all areas of the installation during both day and night?	____	____	____
9. NARCOTICS.			
a) Is the MWD section registered with the Drug Enforcement Administration?	____	____	____
b) Is a *DEA 223 (Controlled Substance Registration Certificate)* on file and current?	____	____	____
c) Have all personnel been designated, in writing, by the PM to handle training aids?	____	____	____
d) Has primary and alternate narcotic training aid custodians been appointed, in writing, by the PM?	____	____	____

Table C-1. Sample Inspection Memorandum (Continued)

OFFICE SYMBOL
SUBJECT: MACOM Technical Assistance Visit or Provost Marshal Inspection

		YES	N/A	NO
e)	Have primary and alternate narcotic custodians been cleared by a favorable Crime Records Center name check?	____	____	____
f)	Is the Crime Records Center name check maintained on file?	____	____	____
g)	Have all personnel authorized to handle or use training aids been briefed at least annually?	____	____	____
h)	Is a written record of the briefing being maintained on file?	____	____	____
i)	Are the monthly/quarterly inventories being conducted by a disinterested E7 or above?	____	____	____
j)	Are the monthly/quarterly inventories being maintained with the controlled substances?	____	____	____
k)	Are the monthly/quarterly inventories being maintained for at least two years?	____	____	____
l)	Are joint inventories being conducted on the change of primary or alternate custodians and is the inventory maintained on file?	____	____	____
m)	Is a daily issue/return green logbook on hand?	____	____	____
n)	Is the property book properly maintained to account for all authorized explosive training aids?	____	____	____
o)	Is an *SF 702 (Security Container Check Sheet)* posted on the safe?	____	____	____
p)	Are *SFs 702* maintained for 30 days?	____	____	____
q)	Is a controlled-substance accountability folder on hand?	____	____	____
r)	Is the controlled-substance accountability folder properly maintained?	____	____	____
10.	**EXPLOSIVES.**			
a)	Have primary and alternate explosive custodians been appointed, in writing, by the PM?	____	____	____
b)	Have primary and alternate custodians been cleared by a favorable Crime Records Center name check?	____	____	____
c)	Are records of the Crime Records Center checks on file?	____	____	____
d)	Are all personnel authorized to receive and handle explosive training aids designated, in writing, by PM?	____	____	____
e)	Is annual explosive safety training being conducted by EOD personnel or other qualified personnel?	____	____	____

Table C-1. Sample Inspection Memorandum (Continued)

OFFICE SYMBOL SUBJECT: MACOM Technical Assistance Visit or Provost Marshal Inspection		YES	N/A	NO
f)	Is a written record of the safety training being maintained on file for a minimum of two years?	____	____	____
g)	Is there an SOP that outlines safety procedures in training areas, transportation of explosives, severe weather conditions, and accidents?	____	____	____
h)	Are the monthly/quarterly inventories being conducted by a disinterested E7 or above?	____	____	____
i)	Are the monthly/quarterly inventories maintained on file for at least two years?	____	____	____
j)	Are all required explosive training aids available to ensure that proficiency standards are met?	____	____	____
k)	Are chlorates being replaced quarterly?	____	____	____
l)	Is each type of explosive being stored in a separate vapor proof container?	____	____	____
m)	Is each vapor proof container properly marked?	____	____	____
n)	Has the vehicle that is used to transport explosive training aids been approved by safety personnel?	____	____	____
o)	Does the installation have an explosive detector training plan that has been approved by the installation commander which lists buildings and facilities authorized for training the PEDD teams?	____	____	____
p)	Is a daily issue/return green logbook on hand?	____	____	____
q)	Is the logbook properly maintained to account for all authorized explosive training aids?	____	____	____
r)	Is a *DA Form 7281-R (Command Oriented Arms, Ammunition, & Explosives Security Screening and Evaluation Record)* on hand for each handler that is authorized to handle or receive explosive training aids?	____	____	____
11.	**TRAINING.**			
a)	Does the KM (or handlers) give training to military police on MWD team functions and operational considerations?	____	____	____
b)	Are the MWDs receiving a minimum of 4 hours of patrol and 4 hours of narcotic/explosives proficiency training each week?	____	____	____
c)	Have all certified narcotic/explosive detector teams received quarterly validation testing by the KM?	____	____	____
d)	Is the minimum proficiency standard for assigned patrol/narcotic detector dogs 90 percent and patrol/explosive detector dogs 95 percent?	____	____	____
e)	Was the dog decertified when his proficiency standard was less than 90 percent and less than 95 percent for more than three consecutive months?	____	____	____

Table C-1. Sample Inspection Memorandum (Continued)

OFFICE SYMBOL
SUBJECT: MACOM Technical Assistance Visit or Provost Marshal Inspection

	YES	N/A	NO
12. RECORDS.			
a) Is the handler's information being recorded on the back of *DD Form 1834*?	____	____	____
b) Have all *DA Forms 2807-R* been completed properly?	____	____	____
c) Have all deficiencies for the MWDs been annotated and explained on DA Forms 2807-R?	____	____	____
d) Is a *DA 3992-R* being maintained to record the MWD's reliability?	____	____	____
e) Does each MWD team have a probable cause folder?	____	____	____

13. THE RESULTS OF THIS INSPECTION ARE—

____ EXCELLENT

____ SATISFACTORY

____ UNSATISFACTORY

14. REMARKS. All areas noted with a deficiency will be annotated in the remarks section. A copy of this technical assistance visit will be provided to OPMG, the MACOM program manager, and the KM. All records of technical assistance visits will be maintained on file at the installation for a minimum of two years after the inspection date with the corrective action memorandum.

15. POINT OF CONTACT. The point of contact for this action is the undersigned at COMM: 555-555-1511 or e-mail address.

(Original Signed)
JEFFREY S. CAPTAIN
LTC, USA
MACOM PM

Appendix D
Training Records

This appendix provides samples and details of how to properly use MWD training records. Records must be maintained for the life of the MWD. At least one year of historical documents must be maintained on file at each kennel for every MWD. Records can be discarded two years after the death, adoption, or euthanasia of the MWD. Nearly all MWDs have a dual purpose (PD and PEDD or PNDD); therefore, every MWD should have a PD and a PEDD/PNDD form on file. Kennels that have internet access can obtain records via the Military Working Dog Management System (MWDMS).

DEPARTMENT OF THE ARMY *FORM 2807-R*

D-1. *DA Form 2807-R (Military Working Dog Training and Utilization Record)* (*Figure D-1,* page D-3) is a standard form used by all MWD personnel. It is used to record the training and utilization of PDs. Use the following instructions when preparing a *DA Form 2807-R*:

ADMINISTRATIVE DATA

D-2. Enter the following information in the appropriate blocks:

- The date.
- The MWD's name and type, tattoo number, and age.
- The handler's name and grade.
- The organization and the location.

TRAINING

D-3. The training section of the form provides complete information on all of the training tasks that must be accomplished for a PD to maintain proficiency. It is not necessary to train on all of the controlled-aggression tasks during the same day; however, all of the controlled-aggression tasks must be trained every week.

D-4. Entries for the amount of time spent training on specific tasks during a particular day of the month are made in minutes. The monthly total for each specific task will be in hours and will be automatically entered in the far right-hand column titled "Total Hours." The handler's or trainer's evaluation of the MWD's daily rating is entered as S or U. Remarks are mandatory for any area the MWD fails to perform to standard. These remarks will be annotated on the reverse side of the form using the following format:

- Deficiency (DF). The MWD's shortcoming.
- Corrective action (CA). The handler's response to the MWD's deficiency.
- Corrective response (CR). The MWD's response to the CA.

D-5. Enter the appropriate information on—

- **Line 1.** Record the number of minutes per day that the MWD is trained on obedience on leash.
- **Line 2.** Record the number of minutes per day that the MWD is trained on obedience off leash.
- **Line 3.** Record the number of minutes per day that the MWD is trained on the obedience course.

- **Line 4.** Record the daily rating of the MWD on controlled aggression (either S or U). Write an explanation of any U rating on the reverse side of the form. In 4a to 4e, record the number of minutes per day that the MWD trained on each controlled-aggression task.
- **Line 5.** Record the number of minutes per day that the MWD trained on building searches on line 5.
- **Line 6.** Record the number of minutes per day that the MWD trained on gunfire with the handler firing the weapon. The number of rounds will be annotated on the reverse side of the form.
- **Line 7.** Record the number of minutes per day that the MWD is trained on gunfire with a third party firing a weapon. The number of rounds will be annotated on the reverse side of the form.
- **Line 8.** Record the number of minutes per day that the MWD is trained on scouting/patrolling. Record the distance (in feet) for scent, sight, and sound detection on lines 8a, b, or c.
- **Line 9.** Record the number of minutes per day that the MWD is trained on vehicle patrol.
- **Line 10.** Record the number of minutes per day that the MWD is trained on tracking.
- **Line 11.** Record the daily training rating (S or U). Explain deficiencies on the reveres side of the form. Any failure of a critical task requires a U daily training rating. See *DA Pam 190-12* for critical tasks.

UTILIZATION

D-6. The utilization section of the form provides a daily record of the time spent performing military police duties in the three general categories of combat support operations, patrol–law enforcement, and patrol–security.

D-7. The combat support operations category may be used to record time spent performing military police missions (actual or training) in support of combat units. Such operations include field training exercises, CP exercises, mobilization exercises, and other activities related to the tactical and strategic missions of the Army, provided that these missions are accomplished with a handler and an MWD together as a certified team.

D-8. The patrol–law enforcement category may be used to record time spent performing law enforcement or force protection type missions. These missions include but are not limited to—

- Random vehicle searches.
- Law enforcement patrols.
- Other duties performed while working in a law enforcement type mission.

D-9. The patrol–security category may be used to record time spent performing security type missions such as—

- Dignitary security.
- Critical asset patrolling in a peacetime environment.
- USSS, Department of State, or other explosive type searches.

D-10. The daily rating of the MWD's performance on military police duties is recorded as S or U. All U performance annotations will have an explanation for the DF and have the CA taken recorded on the reverse side of the form and continuation sheets, if necessary. The total monthly utilization hours are recorded in the last column, titled "Total Hours."

D-11. When training is conducted during the actual use and employment of an MWD, the time used for training may also be reported in the training section. This training is not an automatic entry every time the team has utilization hours. Time annotated as training during utilization hours will be actual time training on patrol or detector tasks. It is unethical to count all or even half of an MWD team's utilization hours as both utilization and training time for every shift, and unrealistic numbers will not be accepted. When time is recorded as utilization and training, a notation is made on the reverse side of the form explaining the double time entry and the type of training conducted. This will make it possible to differentiate between training time, training time while on duty, and duty (utilization) time. Although additional training during utilization hours is acceptable, it will not be accepted as the primary amount of mandatory training hours each week.

MILITARY WORKING DOG TRAINING AND UTILIZATION RECORD For use of this form, see AR 190-12; proponent agency is ODCSOPS						MONTH AND YEAR SEP 2003
NAME/TYPE OF DOG NATZ/PEDD	TATTOO/NUMBER E309	AGE 2	NAME OF HANDLER BOOKSTEIN, DAVID B.	GRADE SGT	ORGANIZATION AND LOCATION HHC 98TH ASG, BAMBERG, GE	

DAILY RATINGS: S - SATISFACTORY, U - UNSATISFACTORY (Explain deficiency and corrective action on reverse)
(Use reverse side of this form for any remarks or notes)

TRAINING	DAY OF MONTH																															TOTAL HOURS
	1	2	3	4	5	6	7	8	9	10	11	12	13	14	15	16	17	18	19	20	21	22	23	24	25	26	27	28	29	30	31	
1. ON LEASH OBEDIENCE	15	15	15	15	30		30	40	10	15			20	15		15	20	30	15		15	15	15	15					15	15		6.50
2. OFF LEASH OBEDIENCE	20	20	15	15	40		30	40	10	15			20	30		15	30	30	15		15	20	20	20					15	15		7.50
3. OBEDIENCE COURSE					20		20			15																						0.92
4. CONTROLLED AGGRESSION (S OR U)		S	S					S	S				S	S																		
a. FALSE RUN		5	5					15	5				5	5																		0.67
b. ATTACK		10	10					20	10				10	10																		1.17
c. SEARCH AND ATTACK		5	5					15	10				10	10																		0.92
d. STAND OFF		10	15					20	5				10	10																		1.17
e. ESCORT		5	5					15	5				5	5																		0.67
5. BUILDING SEARCH		15						15																20								0.83
6. GUNFIRE - HANDLER								10																								0.17
7. GUNFIRE - DECOY/AGITATOR					10														30													0.67
8. SCOUTING/PATROLLING (TIME)					15																			15					15			0.75
a. SCENT DETECTION (DISTANCE)					40																			30					30			
b. SIGHT DETECTION (DISTANCE)																																
c. SOUND DETECTION (DISTANCE)																																
9. VEHICLE PATROL																																0.00
10. TRACKING																																0.00
11. DAILY TRAINING RATING (S OR U)	S	S	S	S	S		S	S	S	S						S	S	S	S		S	S	S	S					S	S		21.92

UTILIZATION	HOURS PER DAY																															TOTAL HOURS
	1	2	3	4	5	6	7	8	9	10	11	12	13	14	15	16	17	18	19	20	21	22	23	24	25	26	27	28	29	30	31	
1. COMBAT SUPPORT OPERATIONS																																0
2. PATROL - LAW ENFORCEMENT	8	8	8	8	8			8	8	8			8	8	8	8				8	8	8	8						8	8		144
3. PATROL - SECURITY																																0
4. DAILY UTILIZATION RATING (S OR U)	S	S	S	S	S			S	S	S			S	S	S	S				S	S	S	S						S	S		144.0

DAILY FEEDING	QUANTITY OF FOOD (BY WEIGHT)																															WT OF DOG	
	1	2	3	4	5	6	7	8	9	10	11	12	13	14	15	16	17	18	19	20	21	22	23	24	25	26	27	28	29	30	31	DATE	LBS
2 CU S/D	A	A	A	A	A	A	A	A	A	A	A	A	A	A	A	A	A	A	A	A	A	A	A	A	A	A	A	A	A	A	A	1ST	75
2 CU S/D	A	A	A	A	A	A	A	A	A	A	A	A	A	A	A	A	A	A	A	A	A	A	A	A	A	A	A	A	A	A	A	15TH	73.5

DA FORM 2807-R, Oct 84 EDITION DEC 72 IS OBSOLETE, AND REPLACES DA FORMS 2810-R AND 2815-R, DEC 72

Sample

Figure D-1. Sample DA Form 2807-R

DA Form 2807-R
Continuation Sheet

DATE	REMARKS
2-Sep-03	**DF1:** MWD WAS DISTRACTED BY OTHER HANDLER; WOULD NOT FOCUS ON COMMANDS.
	CA1: CONDUCTED PROPER AVOIDANCE TRAINING TECHNIQUE.
	CR1: AFTER MWD DID NOT RESPOND TO INITIAL AVOIDANCE TRAINING TECHNIQUE, A PINCH COLLAR WAS USED, RESULTING IN REGAINED FOCUS.
	DF2: MWD BIT SLEEVE ON STAND OFF.
	CA2: CONDUCTED PROPER AVOIDANCE TRAINING TECHNIQUE.
	CR2: MWD PERFORMED STAND OFF CORRECTLY.
5-Sep-03	GUNFIRE TRAINING: 5 ROUNDS FIRED BY THIRD PARTY.
	DF1: MWD REPEATEDLY BROKE POSITION (SIT TO DOWN AND DOWN TO SIT).
	CA1: CONDUCTED PROPER AVOIDANCE TRAINING TECHNIQUE.
	CR1: MWD REMAINED IN POSITION.
	DF2: MWD WAS HESITANT TO OUT REWARD (BITE SLEEVE).
	CA2: CONDUCTED PROPER AVOIDANCE TRAINING TECHNIQUE.
	CR: MWD OUTED REWARD ON COMMAND.
7-Sep-03	MWD HAS BEEN BREAKING POSITION AFTER SHORT TIME DURING OFF/L OB WHEN HANDLER IS POSITIONED AWAY FROM HIM. TODAY MWD REMAINED IN THE SIT FOR 1 MIN 50 SEC, BROKE SIT, WAS VERBALLY CORRECTED AND REMAINED IN SIT FOR AN ADDITIONAL 25 SEC.
	DF: MWD BROKE FROM DOWN SEVERAL TIMES DURING PARADE REST.
	CA: CONDUCTED PROPER AVOIDANCE TRAINING TECHNIQUE.
8-Sep-03	GUNFIRE TRAINING: 10 ROUNDS FIRED BY HANDLER.
	CR: AFTER MWD DID NOT RESPOND TO AVOIDANCE TRAINING, ESCAPE TRAINING WAS CONDUCTED. EVENTUALLY MWD REMAINED IN DOWN. MWD IS SO EXCITABLE THAT HE HAS A TENDENCY TO CUE OFF HANDLER'S MOVEMENTS TO BREAK FROM POSITION. HANDLER IS CONCENTRATING ON TEACHING MWD NOT TO BE SO DISTRACTED BY OUTSIDE STIMULUS.
	DF1: MWD WAS GETTING DISTRACTED AND NOT HEELING IN THE PROPER POSITION.
	CA1: CONDUCTED PROPER AVOIDANCE TRAINING TECHNIQUE (HEEL-A-WAYS).
	CR1: MWD HEELS CORRECTLY.
	DF2: MWD WAS BITING SLEEVE DURING STAND OFF.
11-Sep-03	**CA2:** CONDUCTED PROPER AVOIDANCE TRAINING TECHNIQUE.
	CR2: MWD PERFORMED STAND OFF CORRECTLY.
	NO TRAINING DUE TO MWD GOING TO VET. MWD HAS BEEN DIAGNOSED WITH A POSSIBLE PULLED/STRAINED MUSCLE IN HIS RIGHT LEG, WHICH HE IS FAVORING AS HE WALKS. X-RAYS SHOW NO BREAKS OR FRACTURES. MWD CANNOT DO ANYTHING STRENUOUS FOR THE NEXT WEEK (NO RUNNING OR JUMPING UNTIL 18 SEP) INCLUDING BITE WORK OR OB COURSE. MWD CAN DO BASIC OB STARTING TUES, SEP 16.
19-Sep-03	GUNFIRE TRAINING: 15 ROUNDS FIRED BY DECOY.

Figure D-1. Sample DA Form 2807-R (Continued)

D-12. Record the appropriate utilization information on—

- **Line 1.** Record the hours per day that the MWD is used for combat support operations.
- **Line 2.** Record the hours per day that the MWD is used for patrol–law enforcement.
- **Line 3.** Record the hours per day that the MWD is used for patrol–security.
- **Line 4.** Record the daily utilization rating (S or U). Explain any U on the reverse side of the form.

TOTAL HOURS

D-13. Record the total hours of training and use in the last column on the right side of the form. The time should be recorded in hours and minutes. The MWDMS makes this entry automatically.

DAILY FEEDING

D-14. Record the quantity of food that the MWD eats every day in this section. If the MWD is fed once daily, the type and amount of food will be noted in the first block to the left. Under quantity of food, the amount eaten for each feeding will be annotated as follows:

- "A" if all food was eaten.
- "3/4" if three-quarters was eaten.
- "1/2" if half was eaten.
- "1/4" if one-quarter was eaten.
- "0" if none of the food was eaten.

D-15. If the MWD is fed twice daily, the second block is used to record the second daily feeding (which could be a different type of food). If the MWD is fed once daily, the second block may be used to record a change in the diet of the MWD, directed by the veterinarian. Block 2 may also be used to indicate the type and amount of food used daily for MWDs on a food reward system. Monitoring MWD food consumption is a critical portion of the required daily health checks.

D-16. The last column, titled "Wt of Dog," is used to record a semimonthly weight for the MWD. The weight checks occur on or around the first and the fifteenth day of each month. There is sufficient room under the headings "Date" and "Lbs" to record both the date when the MWD is weighed and the weight in pounds. Any drastic changes in food consumption or weight gain or loss are reported to the veterinarian.

D-17. Enter the following information in the appropriate blocks:

- Enter the VCO's prescribed amount of food to be given daily directly under "Daily Feeding." Both blocks are used when an MWD is fed twice daily.
- Record the amount of food eaten per day, in the appropriate day's block under "Quantity of Food (by Weight)."
- Record the date and weight of the MWD in the respective blocks in the bottom right-hand corner.

REMARKS

D-18. The reverse side of *DA Form 2807-R* (*Figure D-1*) is used to annotate advanced techniques used during training and DFs with CAs and CRs listed. This section is very important for probable cause and shows the time line of events for training. Each day is listed with the action taken on that day. If no training is conducted due to company type training, the words "soldier development" are written on the line followed by the type of training conducted. "Soldier development" is only used when MWD handlers are conducting training in which the MWD is not needed or not taken. If no training is conducted due to leave, passes, or scheduled days off, no entry is needed; *DA Form 6 (Duty Roster)* can be used to verify these days.

DEPARTMENT OF THE ARMY *FORM 3992-R* (FIRST PAGE)

D-19. *DA Form 3992-R (Narcotics or Explosives Detector Dog Training and Utilization Record)* (*Figure D-2*, page D-9) is a standard form used by all MWD personnel. It is used to record the use and training of detector dogs. Use the following instructions when preparing a *DA Form 3992-R:*

ADMINISTRATIVE DATA

D-20. Enter the following information in the appropriate blocks:

- The date.
- The MWD's name and type, tattoo number, and age.
- The handler's name and grade.
- The organization and location.

TRAINING

D-21. The training section of the form is used for recording all training of the detector dog team in the particular detection skill. The detection areas of buildings, containers, and vehicles are listed. There is one blank space for detector training in other areas. Examples of other areas may include—

- Mail facilities.
- Open or wooded areas.
- Warehouses.
- Aircraft.
- Areas considered significant or unique to the command to which the MWD team is assigned.

D-22. Although some of the examples may be included under one of the three prelisted categories, the others should be listed separately. For areas covered under the three general headings, the handler or trainer annotates the specific location under the appropriate day in the remarks (for example, Smith High School, NCO Club, or Corps Auditorium).

D-23. The top line of each training category is divided diagonally into two blocks. The top of each block is used to record the number of training aids planted for that day's detector training. The bottom of each block is used for the number of training aids found by the proper response of the detector dog. If the MWD responds on the same training aid more than once in the same training scenario, it will not be counted as a find unless the number of plants is changed as well. At no time will there be more finds annotated than there are plants. Continually allowing the MWD to return to the last aid found creates false response problems in the MWD and should not be encouraged. It is the handler's responsibility to keep the MWD from returning to the scent cone of the last found training aid, whether the MWD is being worked on or off leash. Once the MWD has entered the scent cone, he must be allowed to work to the source and final respond. At no time will an MWD be yanked off an aid once he is working the scent cone.

D-24. The bottom line of each training category is used to record the search time. Search time is the amount of time devoted to detector training that day. Detector training is the time from the moment the handler enters a problem and commands the MWD to "Seek" until the MWD has completed the search of that training problem, excluding handler or MWD breaks. Add the daily time entries in each training category, and enter the results in the last column, titled "Total Hours."

D-25. Record the following information beside the appropriate training location (building, containers, vehicles, or other):

- **Plants/Finds.** Record the number of plants in the top half of the block and the number of finds in the lower half.
- **Search Time.** Record the amount of time devoted for detector training that day.
- **Training Total Hours.** Record the sum of hours performing detection missions in the far right

UTILIZATION

D-26. All operational detection missions are recorded in the utilization section. The detection missions are listed. Detection missions that are significantly different from these are listed separately to give an accurate record of the types of missions for which the detector dog team is used. There are two blocks under each day of the month for each utilization category. The top block is for recording the number of times the MWD detected a substance. The bottom block is for recording the total search or inspection time for the detection mission. The last column is the sum of the daily times. It gives a record of the total amount of time spent performing the detection missions for that month. The number of finds for the month can also be totaled and entered in the last column, above the time entry.

D-27. Record the following information beside the appropriate location (building, containers, vehicles, or other):

- **Finds.** Record the number of times the MWD finds the substance on a detection mission.
- **Search Time.** Record the total search time.

Detector Dog Search Data

D-28. All relevant information about the productivity and success of each detection mission is recorded using the "Detector Dog Search Data" section. The location may be a building number, a unit designation, map grid coordinates, and/or any other information that helps identify where the detection mission was performed.

D-29. The MPR number is the number assigned to the case by military police to account for the custody and disposition of the substances found. The substance is the identification of the found material by a common name, such as heroin, marijuana, dynamite, or detonating cord.

D-30. Quantity is the measure of the amount of the substance found. Weight, volume, overall dimensions, length, or any other appropriate measure may be used.

D-31. The "Remarks" section may be used for adding any other relevant information about the substance found. This would include a field or laboratory verification of the type of substance, an EOD evaluation of an explosive device or explosive substance as live or inert, or the presence of other hazardous material in or around the substance found (such as razor blades, trip wires, or poisons). Record any information that may be useful in preparing the team for future searches or that may be applicable or useful to other detector dog teams.

D-32. Record the following information in the appropriate boxes:

- **Time.** Record the time the detector dog performed the mission.
- **Date.** Record the date the detector dog performed the mission.
- **Location.** Record the exact location of the detector mission (such as front gate, Fort Sill, Oklahoma; National Airport, Gabon, Africa; or building 442, Fort Bragg, North Carolina).
- **Military Police Report (MPR) Number.** Record the MPR number in this column, if applicable.
- **Substance.** Record the common name of the substance found, if applicable.
- **Quantity.** Record the quantity of the substance found, if applicable.
- **Remarks.** Use this column to record any other relevant information about the substance found (such as a positive response, no response, or a lab verification).

Detector Dog Proficiency (Second Page)

D-33. A correct response occurs when the MWD detects the substance and responds with the proper final response, and the training aid or substance is found in the general location where the MWD has responded. A not-at-source response is a training tool used to help an MWD team move closer to the source location of a training aid plant. The distance away from the source in order for a not-at-source response to be given is six to ten feet.

D-34. A missed response is when the MWD fails to detect and respond to the presence of a training aid with a final response. A handler miss or error is when—

- The handler fails to search or clear an area with the MWD and a training aid plant is missed.
- The handler fails to read an MWD's change of behavior and pulls the MWD away from the area where a training aid is planted and a training aid plant is missed.
- The handler presents an area but fails to ensure that the MWD is sniffing and a training aid plant is missed.

Handler misses or errors are considered missed explosives or narcotics and are counted against the find rate percentages of the MWD team.

D-35. False responses are noted in the second to last section of *DA Form 3992-R*. A false response is when the MWD responds as if he has detected a substance and the handler accepts that response, but no training aid or substance can be found within a reasonable distance of the MWD's final response.

D-36. The detector dog team's proficiency is computed monthly by adding the total number of correct responses ("a" for the month) and adding the total number of false/missed (MWD or handler) responses ("b" for the month). These two numbers are used in the following formula to obtain the detector dog team's proficiency rating:

$$\left(\frac{a}{a+b}\right)100 = proficiency$$

D-37. For example, during the month the detector dog team made 93 correct responses on training aids, the MWD team also had 4 false responses, 2 MWD and 1 handler missed responses for a total of 7 false/missed responses. Applying these two numbers (a=93 and b=7) to the formula, the following is obtained:

$$\left(\frac{93}{93+7}\right)100 = 93\%$$

D-38. The computation above shows that the MWD team is working at a 93 percent proficiency rate. This rate is above the minimum for a PNDD team (90 percent), but it is below the minimum standard for a PEDD team (95 percent). A PNDD handler should continue training to maintain and possibly increase the team's proficiency. A PEDD handler needs to identify the causes of the team's substandard performance and immediately begin corrective training to bring the team up to and over the minimum 95 percent detection proficiency standard.

Note: Responses on actual substances during actual searches are not included in this computation. Search data cannot be included because it is impossible to determine the number of false or missed responses during an actual search.

D-39. Record the following information on—

- **Line a.** Record the total correct alerts. The MWDMS makes these entries automatically.
- **Line b.** Record the total false or missed alerts. The MWDMS makes these entries automatically.

Narcotics or Explosive Detector Dog Training and Utilization Record (Third and Fourth Pages)

D-40. The first part of section of the Narcotics or Explosive Detector Dog Training and Utilization Record section is used to record the number of each training aid used in training on each day of the month. The entry for each training aid type is divided into three rows to record the day of the month, correct alerts, and missed alerts, respectively. Record the appropriate information in the following sections.

- **Administrative Data.** Record the MWD's name and tattoo number, the month, and the year in the appropriate blocks.
- **Aid Type.** Enter the specific aid type in the upper left-hand corner of each "Detector Dog Proficiency" section.
- **Total Correct Alerts.** Record the number of correct alerts the MWD made during training.
- **Missed Alerts.** Record the number of missed and false alerts the MWD made during training.
- **Proficiency.** Record the actual proficiency in this block. The MWDMS makes this entry automatically.

Remarks

D-41. Remarks for *DA Form 3992-R* are recorded on a continuation sheet. Each DF (for example, a missed aid or false sit) is listed by the date it occurred. The CA and CR are also listed. If more than one DF is noted per day, number the DF, CA, and CR series to avoid confusion.

NARCOTICS OR EXPLOSIVES DETECTOR DOG TRAINING AND UTILIZATION RECORD

For use of this form see AR 190-12. The proponent is DCSPER.

MONTH AND YEAR	NAME/TYPE OF DOG	TATTOO NUMBER	AGE	NAME OF HANDLER	GRADE	ORGANIZATION AND LOCATION
SEP 2003	NATZ/PEDD	E309	2	BOOKSTEIN, DAVID B.	SGT	HHC 98TH ASG, BAMBERG, GE

TRAINING		1	2	3	4	5	6	7	8	9	10	11	12	13	14	15	16	17	18	19	20	21	22	23	24	25	26	27	28	29	30	31	TRAINING TOTAL HOURS 24.25
BUILDINGS	PLANTS /			7	8	4				4	6						11		5						4	8	6				7		70
	FINDS			7	5	4				4	6						10		4						4	8	6				7		65
	SEARCH TIME			120	80	60				60	60						130		70						60	120	120				120		16.67 HRS
CONTAINERS	PLANTS /									19	57						19		19														114
	FINDS									19	57						19		19														114
	SEARCH TIME									45	120						45		45														4.25 HRS
VEHICLES	PLANTS /					4													5														9
	FINDS					4													5														9
	SEARCH TIME					60													80														2.33 HRS
OTHER	PLANTS /									4																							4
	FINDS									4																							4
	SEARCH TIME									60																							1.00 HRS

UTILIZATION		1	2	3	4	5	6	7	8	9	10	11	12	13	14	15	16	17	18	19	20	21	22	23	24	25	26	27	28	29	30	31	UTILIZATION TOTAL HOURS 0.00
BUILDINGS	FINDS																																0
	SEARCH TIME																																0.00 HRS
CONTAINERS	FINDS																																0
	SEARCH TIME																																0.00 HRS
VEHICLES	FINDS																																0
	SEARCH TIME																																0.00 HRS
OTHER	FINDS																																0
	SEARCH TIME																																0.00 HRS

DA FORM 3992-R, Oct 84 EDITION OF 1 DEC 72 IS OBSOLETE

Figure D-2. Sample DA Form 3992-R

DETECTOR DOG SEARCH DATA

NO.	TIME	DATE	LOCATION	MPR NUMBER	SUBSTANCE	QUANTITY	REMARKS
1							
2							
3							
4							
5							
6							
7							
8							
9							
10							
11							
12							
13							
14							
15							
16							
17							
18							
19							
20							
21							
22							
23							
24							
25							
26							
27							
28							
29							
30							
31							
32							
33							

DETECTOR DOG PROFICIENCY

(Alerts on training aids during training and actual searches)

DAY OF MONTH	1	2	3	4	5	6	7	8	9	10	11	12	13	14	15	16	17	18	19	20	21	22	23	24	25	26	27	28	29	30	31	PROFICIENCY = a / (a + b) X 100
a. TOTAL CORRECT ALERTS	0	0	7	5	8	0	0	0	26	63	0	0	0	0	0	29	0	28	0	0	0	0	0	4	8	6	0	0	0	7	0	191
b. TOTAL FALSE/MISSED ALERTS	0	0	0	3	0	0	0	0	1	0	0	0	0	0	0	3	0	1	0	0	0	0	0	0	0	0	0	0	0	0	0	199 96%

DA Form 3992-R, Oct 84

Figure D-2. Sample DA Form 3992-R (Continued)

NARCOTICS OR EXPLOSIVES DETECTOR DOG TRAINING AND UTILIZATION RECORD

NATZ/PEDD ______ TATTOO: E309 MONTH: SEP YEAR: 2003

DETONATING CORD

DETECTOR DOG PROFICIENCY (ALERTS ON TRAINING AIDS DURING TRAINING AND ACTUAL SEARCHES)

DAY OF MONTH	1	2	3	4	5	6	7	8	9	10	11	12	13	14	15	16	17	18	19	20	21	22	23	24	25	26	27	28	29	30	31	PROFICIENCY A / A + B X 100
a. TOTAL CORRECT ALERTS			2	1						20						1									3							27
b. TOTAL MISSED ALERTS				1																												28 96%

C-4 PLASTIC EXPLOSIVE

DETECTOR DOG PROFICIENCY (ALERTS ON TRAINING AIDS DURING TRAINING AND ACTUAL SEARCHES)

DAY OF MONTH	1	2	3	4	5	6	7	8	9	10	11	12	13	14	15	16	17	18	19	20	21	22	23	24	25	26	27	28	29	30	31	PROFICIENCY A / A + B X 100
a. TOTAL CORRECT ALERTS				1					22	1						3		2								1						30
b. TOTAL MISSED ALERTS				1					1																							32 94%

POTASSIUM CHLORATE

DETECTOR DOG PROFICIENCY (ALERTS ON TRAINING AIDS DURING TRAINING AND ACTUAL SEARCHES)

DAY OF MONTH	1	2	3	4	5	6	7	8	9	10	11	12	13	14	15	16	17	18	19	20	21	22	23	24	25	26	27	28	29	30	31	PROFICIENCY A / A + B X 100
a. TOTAL CORRECT ALERTS									1							20		2						2	1							26
b. TOTAL MISSED ALERTS																1																27 96%

SODIUM CHLORATE

DETECTOR DOG PROFICIENCY (ALERTS ON TRAINING AIDS DURING TRAINING AND ACTUAL SEARCHES)

DAY OF MONTH	1	2	3	4	5	6	7	8	9	10	11	12	13	14	15	16	17	18	19	20	21	22	23	24	25	26	27	28	29	30	31	PROFICIENCY A / A + B X 100
a. TOTAL CORRECT ALERTS					1				2									21						1	1							26
b. TOTAL MISSED ALERTS																		1														27 96%

WATER GEL (TOVEX)

DETECTOR DOG PROFICIENCY (ALERTS ON TRAINING AIDS DURING TRAINING AND ACTUAL SEARCHES)

DAY OF MONTH	1	2	3	4	5	6	7	8	9	10	11	12	13	14	15	16	17	18	19	20	21	22	23	24	25	26	27	28	29	30	31	PROFICIENCY A / A + B X 100
a. TOTAL CORRECT ALERTS				1						21						1		1								2				2		28
b. TOTAL MISSED ALERTS				1																												29 97%

DA FORM 3992-R, OCT 84 (CONTINUATION SHEET)

Figure D-2. Sample DA Form 3992-R (Continued)

NARCOTICS OR EXPLOSIVES DETECTOR DOG TRAINING AND UTILIZATION RECORD

NATZ/PEDD TATTOO: E309 MONTH: SEP YEAR: 2003

SEMTEX	DETECTOR DOG PROFICIENCY (ALERTS ON TRAINING AIDS DURING TRAINING AND ACTUAL SEARCHES)																															PROFICIENCY A / A + B X 100
DAY OF MONTH	1	2	3	4	5	6	7	8	9	10	11	12	13	14	15	16	17	18	19	20	21	22	23	24	25	26	27	28	29	30	31	A + B
a. TOTAL CORRECT ALERTS			1		2				1							2		2							1					2		11
b. TOTAL MISSED ALERTS																																~~11~~ 100%
AMMONIUM NITRATE	DETECTOR DOG PROFICIENCY (ALERTS ON TRAINING AIDS DURING TRAINING AND ACTUAL SEARCHES)																															PROFICIENCY A / A + B X 100
DAY OF MONTH	1	2	3	4	5	6	7	8	9	10	11	12	13	14	15	16	17	18	19	20	21	22	23	24	25	26	27	28	29	30	31	A + B
a. TOTAL CORRECT ALERTS																																0
b. TOTAL MISSED ALERTS																																0 #DIV/0!
NITROGLYCERIN DYNAMITE	DETECTOR DOG PROFICIENCY (ALERTS ON TRAINING AIDS DURING TRAINING AND ACTUAL SEARCHES)																															PROFICIENCY A / A + B X 100
DAY OF MONTH	1	2	3	4	5	6	7	8	9	10	11	12	13	14	15	16	17	18	19	20	21	22	23	24	25	26	27	28	29	30	31	A + B
a. TOTAL CORRECT ALERTS				2						21															1							24
b. TOTAL MISSED ALERTS																																24 100%
TNT (Trinitrotoluene)	DETECTOR DOG PROFICIENCY (ALERTS ON TRAINING AIDS DURING TRAINING AND ACTUAL SEARCHES)																															PROFICIENCY A / A + B X 100
DAY OF MONTH	1	2	3	4	5	6	7	8	9	10	11	12	13	14	15	16	17	18	19	20	21	22	23	24	25	26	27	28	29	30	31	A + B
a. TOTAL CORRECT ALERTS			3		3											2										1				2		11
b. TOTAL MISSED ALERTS																																11 100%
SMOKELESS POWDER	DETECTOR DOG PROFICIENCY (ALERTS ON TRAINING AIDS DURING TRAINING AND ACTUAL SEARCHES)																															PROFICIENCY A / A + B X 100
DAY OF MONTH	1	2	3	4	5	6	7	8	9	10	11	12	13	14	15	16	17	18	19	20	21	22	23	24	25	26	27	28	29	30	31	A + B
a. TOTAL CORRECT ALERTS			1		2																			1	1	2				1		8
b. TOTAL MISSED ALERTS																																8 100%
Totals from TNG Aids (Cont. Sheet)																																
TOTAL PLANTS	0	0	7	8	8	0	0	0	27	63	0	0	0	0	0	30	0	29	0	0	0	0	0	4	8	6	0	0	0	7	0	197
TOTAL FINDS	0	0	7	5	8	0	0	0	26	63	0	0	0	0	0	29	0	28	0	0	0	0	0	4	8	6	0	0	0	7	0	191 97%
a. TOTAL FALSE ALERTS																2																2 1%
Totals from TNG locations (3992 Front)																																
TOTAL PLANTS	0	0	7	8	8	0	0	0	27	63	0	0	0	0	0	30	0	29	0	0	0	0	0	4	8	6	0	0	0	7	0	197
TOTAL FINDS	0	0	7	5	8	0	0	0	27	63	0	0	0	0	0	29	0	28	0	0	0	0	0	4	8	6	0	0	0	7	0	192 97%

DA FORM 3992-R, OCT 84 (CONTINUATION SHEET)

Figure D-2. Sample DA Form 3992-R (Continued)

DA Form 3992-R
Continuation Sheet

DATE	REMARKS
3-Sep-03	TRAINING AT SCHWEINFURT BOWLING ALLEY. (TRAINING AIDS PLANTED FROM 6 FT H x 5 FT D)
4-Sep-03	TRAINING AT SCHWEINFURT FMO. (TRAINING AIDS PLANTED FROM 8 FT H x 3 FT D)
	DF1: MWD MISSED AID (1 STRAND DET CORD - 1 FT H x 1 FT D) DUE TO HANDLER ERROR.
	CA1: HANDLER TOLD TO WORK A REVERSE PATTERN UNTIL THE TEAM REACHED THE AID LOCATION.
	CR1: MWD FOUND AID AFTER HANDLER MADE PROPER PRESENTATION.
	DF2: MWD MISSED AID. (1/4 BLOCK C4 2 FT H x 1 FT D)
	CA2: MWD BROUGHT BACK FOR RE-PRESENTATION.
	CR2: MWD FOUND AID AFTER HANDLER MADE SECOND PRESENTATION.
	DF3: MWD MISSED AID. (1 STICK NITRO DYNAMITE- 7 FT H x 1 FT D)
	CA3: MWD BROUGHT BACK FOR RE-PRESENTATION.
	CR3: MWD FOUND AID AFTER HANDLER MADE HIGH PRESENTATION AND DRAGGED HAND UPWARD.
	DF4: MWD MISSED AID. (1 STICK WATER GEL - 5 FT H x 1 FT D)
	CA4: MWD BROUGHT BACK FOR RE-PRESENTATION.
	CR4: MWD FOUND AID AFTER HANDLER MADE SECOND PRESENTATION.
	MWD WILL BE PROTOCOLED ON ALL MISSED AIDS AT A LATER DATE DUE TO TIME CONSTRAINTS.
5-Sep-03	TRAINING AT 630TH MILITARY POLICE COMPANY MOTOR POOL. (TRAINING AIDS PLANTED FROM 3 FT H x 7 FT D)
9-Sep-03	TRAINING AT SCHWEINFURT OPEN AREA BY CONN CLUB & OUTDOOR REC. (TRAINING AIDS PLANTED FROM 7 FT H x 4 FT D)
	CA: (REFERENCE DF2, 4 SEP) MWD WAS PROTOCOLED ON 1/2 BLOCK C4.
	CR: COMPLETED 4 HV ASSISTED FOR 4 TRIALS AND 15 CONSECUTIVE UNASSISTED TRIALS.
10-Sep-03	TRAINING AT ANSBACH CMFO WAREHOUSES (TRAINING AIDS PLANTED FROM 4 FT H x 1 FT D).
	CA: (REFERENCE DF2-4, 4 SEP) MWD WAS PROTOCOLED ON 1 STRAND DC, 1 STICK ND, AND 1 STICK WG.
	CR: COMPLETED 4 HV ASSISTED FOR 4 TRIALS AND 15 CONSECUTIVE UNASSISTED TRIALS FOR EACH AID.
11-Sep-03	NO TRAINING DUE TO MWD GOING TO VET. MWD HAS BEEN DIAGNOSED WITH A POSSIBLE PULLED/STRAINED MUSCLE IN HIS RIGHT LEG WHICH HE IS FAVORING AS HE WALKS. X-RAYS SHOW NO BREAKS OR FRACTURES. MWD CANNOT DO ANYTHING STRENUOUS FOR THE NEXT WEEK, BUT IS CLEARED TO WORK DETECTION NEXT TUES (16 SEP).
16-Sep-03	TRAINING AT KITZIGEN TISA WAREHOUSE & LEIGHTON THEATER. (TRAINING AIDS PLANTED FROM 9 FT H x 6 FT D.)
	DF1: MWD MISSED AID. (PC - 1 FT H x 6 FT D)
	CA1: MWD BROUGHT BACK FOR REPRESENTATION. WILL CONDUCT PROTOCOL.
	CR1: MWD RESPONDED ON AID AFTER HANDLER MADE SECOND PRESENTATION. MWD WAS PROTOCOLED ON PC COMPLETING 4 HV ASSISTED FOR 4 TRIALS AND 15 CONSECUTIVE UNASSISTED TRIALS.
	DF2: MWD FALSE SAT IN THEATER LOBBY.
	CA2: MWD WAS ESCAPED FROM AREA AND CONTINUED TO WORK.
	CR2: MWD CONTINUED PATTERN WITHOUT FALSE SITTING.
	DF3: MWD FALSE SAT IN THEATER BACK WALL.
	CA3: MWD WAS ESCAPED FROM AREA AND CONTINUED TO WORK. MWD WILL BE WORKED ON BLANK PROBLEM.
	CR3: MWD CONTINUED PATTERN WITHOUT FALSE SITTING.
18-Sep-03	TRAINING AT SCHWEINFURT STRIP LOT & OUTDOOR REC. (TRAINING AIDS PLANTED FROM 8 FT H x 6 FT D)
	CA1: (REFERENCE DF2, 16 SEP) CONDUCTED BLANK PROBLEM.
	CR1: MWD DID NOT ATTEMPT TO FALSE SIT.
	DF2: MWD MISSED AID. (SC - 7 FT H x 3 FT D)
	CA2: MWD BROUGHT BACK FOR RE-PRESENTATION. MWD WAS PROTOCOLED ON SC.
	CR2: MWD FOUND AID AFTER HANDLER MADE SECOND PRESENTATION. COMPLETED 4 HV ASSISTED FOR 4 TRIALS AND 15 CONSECUTIVE UNASSISTED TRIALS.
24-Sep-03	TRAINING AT SCHWEINFURT FMO. (TRAINING AIDS PLANTED FROM 5 FT H x 5 FT D)
25-Sep-03	TRAINING AT THEATER. (TRAINING AIDS PLANTED FROM 2 FT H x 2 FT D)
26-Sep-03	TRAINING AT COMMISSARY. (TRAINING AIDS PLANTED FROM 9 FT H x 6 FT D)
30-Sep-03	TRAINING AT FURNITURE WAREHOUSE. (TRAINING AIDS PLANTED FROM 6 FT H x 7 FT D)

Figure D-2. Sample DA Form 3992-R (Continued)

Appendix E

Deployment Equipment List

Table E-1 is a list to be used as a general guide to aid leaders in the planning process prior to MWD deployments. Mission, enemy, terrain, troops, time available, and civilian considerations (METT-TC) should be considered when determining what equipment is necessary for mission success. Required and authorized equipment is determined by tables of organization and equipment (TOEs), modification tables of organization and equipment (MTOEs), and tables of distribution and allowances (TDAs).

Table E-1. Deployment Equipment List

Worn or Carried		
M9 with LCE (LBV)	Identification card	Body armor
Canteen cup	Identification tag	Black gloves with liners
First aid pouch with dressing	Protective mask with a carrier	Military police brassard (subdued)
Kevlar® helmet		
MWD Equipment		
Choke chain	Grooming brush/comb	Feed pan
Leather or nylon collar	Reward ball/Kong	Kennel crate
Stake-out chain	360-inch nylon leash	16-foot leash
Plastic or leather muzzle	Water bucket	Harness
Alice Pack		
BDU (2 sets)	Towels (2)	Sleeping bag with a cover
Brown T-shirt (2)	Gloves, black, cold weather	Shower shoes
Undergarments (2)	BDU Cap	PT uniform (complete set)
Green and black socks (2)	Boot polish kit	MOPP gear in an NBC bag
Poncho with a liner	Flashlight	Boots
Shaving kit	Lip balm	Personal hygiene gear
Waterproof bag	Foot powder	
A/B Bag		
Duffel bag with a padlock	Brown T-shirt (5)	White socks (3)
Cold weather boots	BDU (2)	Green and black socks (5)
Cold weather gear	Towels (3)	Personal hygiene gear
Waterproof bag	Overboots	Undergarments (5)
Military police brassard	Laundry bag (2)	
Additional Equipment		
30-90 day supply MWD food	50-gallon trash can (MWD food)	5-gallon water can
MWD training records	Medical supplies	Engineer tape
MWD medical records	Carpenter kit/tools	Cots
MWD health certificate	Safe for narcotics aides	Fire extinguishers

Table E-1. Deployment Equipment List (Continued)

Additional Equipment		
MWD first aid kit	Explosives and narcotics training aides	Axes, shovels, rakes, brooms
Arm protector, sleeve (hidden)	Parasite preventative mediation	Cooling vest (MWD)
Soldier first aid kit	Light set	Paw pad protectors (2 sets)
Bite suit (optional)	Generator	Foot lockers (2) or blood boxes
Dog shampoo	Transformer	Equipment duffel bag
Heaters and fans	Warning signs (MWD)	Camouflage netting with poles

Appendix F

Force Protection Conditions

The FPCONs outlined below describe the progressive level of a terrorist threat to all US military facilities and personnel under *DOD Directive 2000.12*. As approved by the Joint Chiefs of Staff, the terminology and definitions in this appendix are recommended security measures designed to ease interservice coordination and to support US military antiterrorism activities. Each installation must localize these measures for its specific circumstances, accounting for service and commander requirements, local laws, and SOFAs as a minimum. See *AR 525-13* for further details concerning FPCONs. This appendix details what each condition is and the suggested actions of the MWD teams for each level and measure. Commanders are encouraged to employ MWD teams wherever an MWD team can enhance security and force protection measures that are already established.

FORCE PROTECTION CONDITION NORMAL

F-1. FPCON normal applies when a general global threat of possible terrorist activity exists. It warrants a routine security posture.

FORCE PROTECTION CONDITION ALPHA

F-2. FPCON Alpha applies when there is an increased general threat of possible terrorist activity against personnel or facilities, the nature and extent of which are unpredictable. Alpha measures must be capable of being maintained indefinitely. The following bullets explain the FPCON Alpha measures:

- **Alpha measure 1.** Inform personnel and family members of the general situation at regular intervals. Also, remind them to be alert for and report suspicious activities such as the presence of unfamiliar personnel and vehicles, suspicious/unattended parcels, and possible surveillance attempts.
- **Alpha measure 2.** Ensure that the duty officer or personnel with access to building plans as well as the plans for area evacuations are available at all times. Plans to execute access control procedures should be in place. Key personnel required to implement security plans should be on call and readily available.
- **Alpha measure 3.** Secure and randomly inspect buildings, rooms, and storage areas not in regular use.

Note: MWD should randomly inspect buildings after normal duty hours. The KM should ensure that no inspection time or location patterns are ever set by MWD teams.

- **Alpha measure 4.** Increase security spot checks of vehicles and persons entering installations under the jurisdiction of the United States.

Note: MWD should perform vehicle checks randomly at entry points.

- **Alpha measure 5.** Limit access points for vehicles and personnel commensurate with a reasonable flow of traffic.

- **Alpha measure 6.** Apply Alpha measures 4, 5, 7, or 8 either individually or in combination with each other as a deterrent.
- **Alpha measure 7.** Review intelligence, counterintelligence, and operations dissemination procedures.
- **Alpha measure 8.** Review and, if necessary, implement security measures for HRP.
- **Alpha measure 9.** Consult local authorities on the threat and mutual antiterrorism measures as appropriate.
- **Alpha measure 10.** Review all plans and be prepared to implement a higher FPCON.
- **Alpha measure 11 (spare).** Add additional measures as deemed appropriate by the commanders.

FORCE PROTECTION CONDITION BRAVO

F-3. FPCON Bravo applies when an increased or more predictable threat of terrorist activity exists. Sustaining Bravo measures for a prolonged period may affect operational capabilities and impact relations with local authorities. The following bullets explain the FPCON Bravo measures:

- **Bravo measure 1.** Continue, or introduce, all measures in the previous FPCON level.
- **Bravo measure 2.** Check plans for the implementation of the next FPCON level.
- **Bravo measure 3.** Identify critical and high-occupancy buildings. Move cars and objects (such as crates and trash containers) away from buildings to reduce vulnerability to bomb attacks. Consider centralized parking.
- **Bravo measure 4.** Secure and inspect all buildings, rooms, and storage areas not in regular use.
- **Bravo measure 5.** Inspect the interior and exterior of buildings in regular use for suspicious packages at the beginning and end of each workday, as well as at random intervals. Patrol teams search the exterior of high-occupancy buildings at the beginning and end of each day.

Note: Use PEDD to search buildings and areas.

- **Bravo measure 6.** Implement mail-screening procedures to identify suspicious letters and parcels.

Note: Search the mail distribution center with both PEDD and PNDD teams randomly.

- **Bravo measure 7.** Inspect commercial deliveries randomly. Advise family members to check home deliveries.
- **Bravo measure 8.** Increase presence around and surveillance of domestic accommodations, schools, messes, clubs, and other soft targets to improve deterrence and defense and build confidence among staff and dependents.

Note: Randomly patrol and perform walking patrols of populated areas with MWDs.

- **Bravo measure 9.** Implement plans to enhance off-installation security of DOD facilities. In areas with threat levels of moderate, significant, or high, coverage includes facilities (such as DOD schools and daycare centers) and transportation services and routes (such as bus routes) used by DOD employees and family members.
- **Bravo measure 10.** Inform local security committees of actions being taken.
- **Bravo measure 11.** Verify the identity of visitors and randomly inspect their suitcases, parcels, and other containers.
- **Bravo measure 12.** Conduct random patrols to check vehicles, people, and buildings.
- **Bravo measure 13.** Implement additional security measures for HRP.
- **Bravo measure 14.** Identify and brief personnel who may augment guard forces. Review specific rules of engagement including the use of deadly force.

- **Bravo measure 15.** Test the mass notification system.
- **Bravo measure 16.** Verify the identity of all personnel entering buildings as deemed appropriate.
- **Bravo measure 17.** Identify off-installation safe havens for use at the next FPCON.
- **Bravo measure 18.** Review and implement proper OPSEC, communications security (COMSEC), and information security (INFOSEC) status.
- **Bravo measure 19.** Consider using tactical deception techniques.
- **Bravo measure 20 (spare).** Add additional measures as deemed appropriate by the commanders.

FORCE PROTECTION CONDITION CHARLIE

F-4. FPCON Charlie applies when an incident occurs or intelligence is received indicating that some form of terrorist action or targeting against personnel or facilities is likely. Implementing FPCON Charlie measures creates hardship and affects the activities of the unit and its personnel. The following bullets explain the FPCON Charlie measures:

- **Charlie measure 1.** Continue, or introduce, all measures in the previous FPCON.
- **Charlie measure 2.** Ensure that duty personnel are cognizant of actions required to implement antiterrorism and force protection plans. Recall required personnel.
- **Charlie measure 3.** Reduce access control to a minimum and strictly enforce entry. Randomly search vehicles.

Note: Increase the PEDD and PNDD teams' search times at entry points.

- **Charlie measure 4.** Increase standoff from sensitive buildings based on the threat.
- **Charlie measure 5.** Issue weapons and appropriate communication equipment to guards according to SOFAs. Include specific orders on the issue of ammunition.
- **Charlie measure 6.** Increase patrols of the installation to include waterside perimeters, if appropriate.

Note: Use MWD assets to expand the perimeter patrol area and detect intruders.

- **Charlie measure 7.** Implement a barrier plan to prevent a vehicle-borne attack.
- **Charlie measure 8.** Protect all designated vulnerable points. Give special attention to vulnerable points outside the military establishment.
- **Charlie measure 9.** Consult local authorities about closing public (and military) roads and facilities to reduce the vulnerability to attack.
- **Charlie measure 10.** Consider searching suitcases, briefcases, and packages being brought onto the installation through ACPs and consider randomly searching suitcases, briefcases, and packages leaving.

Note: Use PEDD and PNDD teams to search the items above.

- **Charlie measure 11.** Review personnel policy procedures to determine a course of action for family members.
- **Charlie measure 12.** Review access procedures for all non-US personnel and adjust as appropriate.
- **Charlie measure 13.** Consider escorting children to and from DOD schools.
- **Charlie measure 14 (spare).** Add additional measures as deemed appropriate by the commanders.

FORCE PROTECTION CONDITION DELTA

F-5. FPCON Delta applies in the immediate area where a terrorist attack has occurred or when intelligence has been received that terrorist action against a specific location or person is imminent. Normally, FPCON Delta is declared as a localized condition. The following bullets explain the FPCON Delta measures:

- **Delta measure 1.** Continue, or introduce, all measures in the previous FPCON.
- **Delta measure 2**. Augment guards as necessary.
- **Delta measure 3.** Identify all vehicles within operational or mission support areas.
- **Delta measure 4.** Search all vehicles and contents before allowing entrance to the installation.

Note: Use PEDD and PNDD as needed to search vehicles and contents.

- **Delta measure 5.** Control access and implement positive identification of all personnel with no exceptions.
- **Delta measure 6.** Search all suitcases, briefcases, and packages brought into the installation.

Note: Use PEDD and PNDD as needed to search the items above.

- **Delta measure 7.** Escort children to and from DOD schools or close the schools.
- **Delta measure 8.** Make frequent checks of the exterior of buildings and of parking areas. Use MWD teams to make frequent checks of the exterior of buildings and parking areas.
- **Delta measure 9.** Restrict all nonessential movement.
- **Delta measure 10.** Close public and military roads and facilities if permitted.
- **Delta measure (spare).** Add additional measures as deemed appropriate by the commanders.

Appendix G
After-Action Reviews

The AAR process is used to document and discuss MWD operations and training. AARs are focused on performance standards that enable other MWD handlers, training NCOs, KMs, program managers, and concerned agencies to discover what happened, why it happened, and how to sustain strengths and improve on weaknesses. Leaders, planners, and developers can use AARs to get the maximum benefit from every mission or task conducted during predeployment, deployment, and redeployment.

AFTER-ACTION REVIEW PURPOSE

G-1. AARs must provide—

- Candid insights into specific soldier, leader, and MWD strengths and weaknesses from various perspectives.
- Feedback and insight critical to future successes.
- Details that evaluation reports alone often lack.

G-2. The USAMPS MWD Career Management NCO collects historical data, end-of-tour reports, and other organizational specific issues. With the submission of AARs, USAMPS can build a historical database to benefit all in the improvement of Army MWD team tactics, techniques, and procedures. Ultimately, information will be submitted to the Center for Army Lessons Learned (CALL) for distribution to the Army.

AFTER-ACTION REVIEW SUBMISSION

G-3. After the completion of TDY missions, special operations, deployments or other significant events, MWD handlers submit AARs to their respective MACOMs. MACOMs forward the AARs to HQDA Office of the Provost Marshal General (OPMG) and USAMPS.

POINT OF CONTACT

G-4. The point of contact for the MWD Career Management NCO is commercial: (573) 563-8040 and defense switched network (DSN): 676-8040/7946. Send hard-copy correspondence to: US Army Military Police School, ATTN: ATSJ-P, 401 MANSCEN Loop, Suite 1058, Fort Leonard Wood, MO 65473-8926.

EXAMPLE AFTER-ACTION REVIEW

G-5. *Figure G-1*, page G-2, should be used as the format for AARs submitted to the USAMPS MWD Career Management NCO. Refer to *TC 25-20* for comprehensive guidance on AARs.

DEPARTMENT OF THE ARMY
42D Military Police Detachment
Ft. Bragg, NC 27380

OFFICE SYMBOL DATE

MEMORANDUM FOR RECORD

SUBJECT: After-Action Review, Task Force Iron Dog, Sep 03 – Nov 03

1. State what was supposed to happen: This paragraph contains a brief introduction of the mission that was being performed by the MWD team or section. Include operation objectives, the overall mission, and the intent from the task force commander as well as any other relevant information.

2. State what happened: This paragraph contains pertinent information about what actually happened during the mission. Was the mission accomplished as stated above or were other objectives established inside of the area of operation?

3. Determine what went right and wrong during the mission:

a. Topic: What was right and wrong with what happened during the mission.

b. Discussion: Give the who, what, when, where, and how for the mission. Give a good summary of what the issue was.

c. Recommendation: Give recommendations to improve the issues that were identified as needing improvement.

4. Point of contact for this memorandum is the undersigned at DSN: 123-4567 or jon.doe@us.army.mil.

Jon Doe
SSG, USA
Kennel Master / PEDD Handler

Note: If you have a positive topic just use Topic and Discussion sections.

Figure G-1. Sample AAR Format

Appendix H

Health and Welfare Inspections Briefing Guide

Commanders frequently request MWD support during health and welfare inspections. Commanders should be made aware that the conduct and responsibility of health and welfare inspections lies within their purview and that the MWD team is not the authority for the mission. This appendix identifies the actions that need to be addressed with the commander to ensure a successful and lawful health and welfare inspection. These actions are needed to establish and act on probable cause determinations. Questions concerning these matters should be directed to the local Staff Judge Advocate Office for clarification.

COORDINATION

H-1. The KM coordinates and approves each health and welfare inspection. After the KM receives the request for assistance, he assesses the needs of the mission and determines which detector dog team (if any) will assist the commander. The KM notifies the unit commander or 1SG and accepts or denies the mission. Once the mission is accepted, the KM briefs them on their roles and responsibilities and informs them that if his criteria are not met, the KM will terminate the search.

HANDLERS BRIEFING

H-2. Prior to the inspection, the KM, plans NCO, or handler ensures that the commander requesting support is aware that the building being inspected must be unoccupied and that all doors are unlocked (or keys available on site). On the arrival of the MWD team at the location of the health and welfare inspection, the handler performs a commander's briefing and demonstration. The handler—

- Explains the probable cause folder to the commander.
- Shows the commander the training aids used for detection.
- Demonstrates the dog's ability to detect.

H-3. A briefing document shows handlers how to record the details of each health and welfare inspection (*Figure H-1*, page H-3). The commander is provided with a copy of the briefing document containing the following information:

- The MWD teams used for the inspection.
- The canines' names and tattoo numbers.
- The scope of the proficiency training of the MWD teams. Discuss with the commander the standards and scope of the detector training required to maintain the certification of assigned MWD teams.
- The MWDs' accuracy rates. Explain to the commander that the accuracy percentage rate is based on quarterly training.
- The MWDs' detection experience. Inform the commander of the amount of time that the MWDs have actually spent conducting detection missions. Include any possible finds in prior inspections.
- The MWDs' capabilities. Explain to the commander that the MWDs are capable of detecting either drugs or explosives.

- The MWDs' alert response. Describe what the MWDs will do when they indicate the presence or odor of found drugs or explosives.
- The unit representatives. Notify the commander that one SSG or above is required to escort each MWD team through out the inspection. The commander may not be a unit representative.

UNIT REPRESENTATIVE DUTIES AND RESPONSIBILITIES

H-4. Ensure that the unit representatives are informed that they—

- Escort the MWD team at all times.
- Unlock doors or areas as needed.
- Carefully observe the MWD team.
- Follow the handler's instructions.
- Obey safety precautions.
- Spot areas that are missed by the MWD team.

DOCUMENTATION

H-5. It is important that the MWD handler documents the commander's briefing on every occasion. *Figure H-1*, page H-3, gives an example for documenting the briefing given to the commander before a health and welfare inspection. Advise the commander to maintain the handlers briefing and other documentation concerning each response of the MWD. Advise the commander that a positive response by a narcotic detector dog provides probable cause for a command-directed urinalysis. Ensure that the commander signs the briefing form and then provide a copy for filing in case of a judicial action.

INSPECTION

H-6. After the briefing has been conducted, the inspection may begin. Remember to advise the commander and unit representatives that this health and welfare inspection is the unit's mission and that the MWD team is assisting in the mission. The MWD team is not in charge of the health and welfare inspection but can be used as a tool by the commander.

ATZP-PM-K DATE: ____________

SUBJECT: Commander's Briefing for Health and Welfare Inspections Utilizing Detector Dogs

1. (commander's name) has requested narcotics detector dog support for a health and welfare inspection on (dd/mm/yyyy) .

2. Prior to canine(s) being utilized, the building must be unoccupied and all doors must be unlocked (or keys available).

3. The following provides the commander with information essential to a successful inspection:

_____ A. Team(s) to be utilized: (number of PEDD and PNDD to be used) .

_____ B. MWD name(s) and tattoo number(s): (name and tattoo number for each MWD)

_____ C. The scope of formal training of the canine team(s). Discuss detector training. Discuss the standards that must be successfully completed to maintain certification of detector dog team(s).

_____ D. The canine's accuracy rate(s) is/are (accuracy rates for each MWD) . This is based on the previous 3 months training records.

_____ E. The extent of actual MWD detection field experience. (experience for each MWD)

_____ F. Explain the types of drugs or explosives that the canine(s) is/are trained to detect.

_____ G. Describe the canine's specific behavior(s)/actions when it responds, indicating the presence or the odor of narcotics (detector response of each MWD) .

_____ H. Ensure that the commander(s) is/are aware that one representatives (SSG or above) is needed to act as an escort for each team. The commander may not be one of the individuals accompanying canine team(s).

_____ I. Ensure that the escorts are briefed on their duties and responsibilities. Discuss the safety precautions that they must follow, and ensure the safety of others.

_____ J. Ensure that the commander(s) is/are aware that the unit is responsible for maintaining proper documentation (date, time, room number, common or personal area and occupant[s] name[s]), concerning the response of the canine(s). This information should be kept on file with the commander's copy of this briefing sheet, in case of a judicial action.

_____ K. Ensure that the commander is aware that a positive response by a narcotic detector dog is probable cause for a command-directed urinalysis on the occupant or occupants of that room.

I (commander's name) have received a personal briefing covering all the information stated above and now understand the capabilities of the canine team(s). I am satisfied that the canine team(s) have the training and experience necessary to provide reliable information upon which to authorize a search and/or perform a command-directed urinalysis. I authorize the canine team(s) to inspect the following area(s): (areas to be inspected)

Figure H-1. Health and Welfare Briefing Document Sample

HANDLER(S) SIGNATURE

1. ______________________ 2. ______________________

3. ______________________ 4. ______________________

COMMANDER'S INFORMATION	**1SG'S INFORMATION**
NAME: ______________	NAME: ______________
RANK: ______________	RANK: ______________
UNIT: ______________	UNIT: ______________
SIGNATURE: ______________	SIGNATURE: ______________
UNIT REPRESENTATIVE	**UNIT REPRESENTATIVE**
NAME: ______________	NAME: ______________
RANK: ______________	RANK: ______________
UNIT: ______________	UNIT: ______________
SIGNATURE: ______________	SIGNATURE: ______________

Figure H-1. Health and Welfare Briefing Document Sample (Continued)

Appendix I

Terrorist Bomb Threat Stand-off Distances

Table I-1 contains information from the DOD technical support working group. Commanders, PMs, and MWD planners can use this information as a guide to determine the minimum distances necessary to evacuate personnel from suspected explosive threats.

Table I-1. Minimum Distances for Personnel Evacuation

Threat Description	*Explosives Capacity* (TNT Equivalent)*	*Building Evacuation Distance***	*Outdoor Evacuation Distance****
Pipe bomb	5 lbs/2.3 kg	70 ft/21 m	850 ft/259 m
Briefcase/suitcase bomb	50 lbs/23 kg	150 ft/46 m	1,850 ft/564 m
Compact sedan	500 lbs/227 kg	320 ft/98 m	1,500 ft/457 m
Sedan	1,000 lbs/454 kg	400 ft/122 m	1,750 ft/534 m
Passenger/cargo van	4,000 lbs/1,814 kg	640 ft/195 m	2,750 ft/838 m
Small moving van/delivery truck	10,000 lbs/4,536 kg	860 ft/263 m	3,750 ft/1,143 m
Moving van/water truck	30,000 lbs/13,608 kg	1,240 ft/375 m	6,500 ft/1,982 m
Semitrailer	60,000 lbs/27,216 kg	1,570 ft/475 m	7,000 ft/2,134 m

* Based on maximum volume or weight of explosives (TNT equivalent).

** Governed by the ability of an unstrengthened building to withstand severe damage or collapse.

*** Governed by either the fragment throw distance or glass breakage/falling glass hazard distance (whichever throws fragments farther than vehicle bombs).

NOTES:
1. All personnel must either seek shelter inside a building away from windows and exterior walls (with some risk) or move beyond the outdoor evacuation distance.
2. The preferred area is beyond the outdoor evacuation distance for the evacuation of personnel in buildings. This is mandatory for personnel outdoors.

Glossary

Acronyms and Abbreviations

1SG	first sergeant
AAR	after-action review
ACP	access control point
AF	Air Force
agg	aggression
AR	Army regulation
ASG	area support group
ASI	additional skill identifier
ASP	ammunition supply point
ATTN	attention
BDU	battle dress uniform
C2	command and control
C4	composition 4
CA	corrective action
CALL	Center for Army Lessons Learned
CCTV	Closed-circuit television
cert	certification
CI	civilian internee
CID	Criminal Investigation Division
clg	cleaning
CMFO	centralized furnishings management office
comm	communications
COMSEC	communications security
CONEX	container express
cont	continuted
CP	command post
CPT	captain
CR	corrective response
DA	Department of the Army
d	deep
DARE	Drug Abuse Resistance Education
DCSPER	Dputy Chief of Stagg for Personnel
DC	detonating cord
DD	Department of Defense
DEA	Drug Enforcement Administration

Dec	December
det	detonating
DF	deficiency
DOD	Department of Defense
DODMWDVS	Department of Defense Military Working Dog Veterinary Services
DSN	defense switched network
DST	drug suppression team
EAC	echelon above corps
EOD	explosive ordnance disposal
EPW	enemy prisoner of war
eval	evaluaiton
FM	field manual
FMO	Furnishing managemetn office
FPCON	force protection condition
ft	foot; feet
GE	Germany
GOV	government owned vehicle
GSA	General Services Administration
h	high
HF	hole, variable
HRP	high-risk personnel
HRT	high-risk target
HVAC	heating, ventilation, and air conditioning
I/R	internment and resettlement
IDS	Intrusion Detection System
IED	improvised explosive device
INFOSEC	information security
IPB	intelligence preparation of the battlefield
iso	isolation
kg	kilogram
KM	kennel master
LAFB	Lackland Air Force base
LAN	local area network
LBS	pounds
LBV	load-bearing vest
LCE	load-carrying equipment
LED	light-emitting diode
LIN	line item number
LTC	lieutenent colonel
m	meter

MACOM	major Army command
MANSCEN	Maneuver Support Center
MCI	military customs inspector
MCIPM	military customs inspector program manager
MCM	Manual for Courts-Martial
mech	mechanical
METL	mission-essential task list
METT-TC	mission, enemy, terrain, troops, time available, and civilian considerations
MEVA	mission-essential vulnerable areas
MILSTRIP	military standard transportation and issue procedures
min	minimum
MMSO	maneuver mobility support operations
MOPP	mission-oriented protective posture
MPR	military police report
MSR	main supply route
MTOE	modification table of organization and equipment
MWD	military working dog
MWDMS	Military Working Dog Management System
NBC	nuclear, biological, and chemical
NCO	noncommissioned officer
NCOIC	noncommissioned officer in charge
no	number
Nov	November
NSN	national stock number
OB	obstacle
Oct	October
ODCSOPS	Office of the Deputy Chief of Staff for Operaitons and Plans
off/L	off leash
on/L	on leash
OIF	Operation Iraqi Freedom
OPMG	Office of the Provost Marshal General
OPSEC	operational security
pam	pamphlet
PC	potassium chlorate
PD	patrol dog
PEDD	patrol explosive detector dog
PIO	police intelligence operations
PM	provost marshal
PMO	provost marshal office

PNDD	patrol narcotic detector dog
POV	privately owned vehicle
PT	physical training
prep	preparation
RAMP	random antiterrorism measures program
rec	recreation
S2	intelligence section
SC	sodium cholorate
sec	seconds
Sep	September
SF	standard form
SFC	sergeant first class
SGT	sergeant
SOFA	Status of Forces Agreement
SOP	standard operating procedures
SP	smokeless powder
SRT	special-reaction team
SSG	staff sergeant
STX	Semtex
TDA	table of distribution and allowancees
TDY	temporary duty
TISA	troop issue subsistence activity
TMP	transportation motor pool
TNT	trinitrotoluene
TOE	table of organization and equipment
TRADOC	United States Army Training and Doctrine Command
Tues	Tuesday
US	United States
USA	United States of America
USC	United States Code
USCS	United States Customs Service
USSS	United States Secret Service
v	version
VCO	veterinary corps officer
vet	veterinarian
VIP	very important person
VTF	Veterinary Treatment Facility
VTR	Veterinary Treatment Record
WT	weight

References

DA Forms are available on the Army Publishing Directorate web site <http://www.apd.army.mil>. DD forms are available from the Office of the Secretary of Defense web site <http://www.dtic.mil/whs/directives/infomgt/forms/formsprogram.htm>.

SOURCES USED

These are the sources quoted or paraphrased in this publication.

AR 190-12. *Military Working Dogs.* 30 September 1993.

AR 190-14. *Carrying of Firearms and Use of Force for Law Enforcement and Security Duties.* 12 March 1993.

DA Pam 190-12. *Military Working Dog Program.* 30 September 1993.

DA Pam 611-21. *Military Occupational Classification and Structure.* 31 March 1999.

DOCUMENTS NEEDED

These documents must be available to the intended users of this publication.

AR 700-81. *DOD Dog Program.* 5 May 1971.

DA Form 2807-R. *Military Working Dog Training and Utilization Record.*

DA Form 3992–R. *Narcotics or Explosives Detector Dog Training and Utilization Record.*

FM 3-19.15. *Civil Disturbance Operations.* 18 April 2005.

READINGS RECOMMENDED

These sources contain relevant supplemental information.

AF Form 1256. *Certificate of Training.*

AR 190-47. *The Army Corrections System.* 5 April 2004.

AR 190-8. *Enemy Prisoners of War, Retained Personnel, Civilian Internees and Other Detainees.* 1 October 1997.

AR 350-1. *Army Training and Education.* 9 April 2003.

AR 40-1. *Composition, Mission, and Functions of the Army Medical Department.* 1 July 1983.

AR 40-3. *Medical, Dental, and Veterinary Care.* 12 November 2002.

AR 40-905. *Veterinary Health Services.* 16 August 1994.

AR 525-13. *Antiterrorism.* 4 January 2002.

AR 600-8-101. *Personnel Processing (In-, Out-, Soldier Readiness, Mobilization, and Deployment Processing).* 15 July 2003.

AR 600-9. *The Army Weight Control Program.* 10 June 1987.

AR 614-200. *Enlisted Assignments and Utilization Management.* 30 September 2004.

DA Form 2028. *Recommended Changes to Publications and Blank Forms.*

DA Form 3161. *Request for Issue or Turn-In.*

DA Form 5513-R. *Key Control Register and Inventory.*

DA Form 6. *Duty Roster*

DA Form 7281-R. *Command Oriented Arms, Ammunition, & Explosives Security Screening and Evaluation Record.*

DA Pam 710-2-1. *Using Unit Supply System (Manual Procedures).* 31 December 1997.

DD Form 1834. *Military Working Dog Service Record.*

DEA Form 223. *Controlled Substance Registration Certificate.*

DOD Directive 2000.12. *DoD Antiterrorism (AT) Program.* 18 August 2003.

DOD Directive 5030.49. *DoD Customs and Border Clearance Program.* 4 May 2004.

DOD Directive 6400.4. *DoD Veterinary Services Program.* 22 August 2003.

DOD Instruction 3025.12. *Military Assistance for Civil Disturbances (MACDIS).* 4 February 1994

FM 21-20. *Physical Fitness Training.* 30 September 1992.

FM 27-10. *The Law of Land Warfare.* 18 July 1956.

FM 3-19.40. *Military Policy Internment/Resettlement Operations.* 1 August 2001.

Lackland AFB Form 375. *Patrol Dog Certification.*

Lackland AFB Form 375a. *Detector Dog Certification.*

MCM, Part III Military Rules of Evidence; Rule 311, *Evidence obtained from unlawful searches and seizures.* 2000.

MCM, Part III Military Rules of Evidence; Rule 313, *Inspections and inventories in the armed forces.* 2000.

MCM, Part III Military Rules of Evidence; Rule 314, *Searches not requiring probable cause.* 2000.

MCM, Part III Military Rules of Evidence; Rule 315, *Probable cause searches.* 2000.

MCM, Part III Military Rules of Evidence; Rule 315(f)(2), *Probable cause determination.* 2000.

MCM, Part III Military Rules of Evidence; Rule 316, *Seizures.* 2000.

SF 702. *Security Container Check Sheet.*

TC 19-210. *Access Control Handbook.* 4 October 2004.

TC 25-20. *A Leader's Guide to After Action Reviews.* 30 September 1993.

USC Title 18, Crimes and Criminal Procedure; Part I, Crimes; Section 1385, *Use of Army and Air Force as posse comitatus.*

Index